INVENTING THE EARTH

Dedication

To Dick Chorley, Mark Melton and David Stoddart.

Chute de la rivière des roches, dans l'île de Bourbon.

Frontispiece *Fire and water. Waterfall over lava, Ile de Bourbon (Mauritius). (Breislak, 1818, plate LII.)*

INVENTING THE EARTH

Ideas on Landscape Development since 1740

Barbara A. Kennedy
Emeritus Fellow of St Hugh's College, Oxford

Blackwell
Publishing

BLACKWELL PUBLISHING
350 Main Street, Malden, MA 02148-5020, USA
9600 Garsington Road, Oxford OX4 2DQ, UK
550 Swanston Street, Carlton, Victoria 3053, Australia

First published 2006 by Blackwell Publishing Ltd

1 2006

Library of Congress Cataloging-in-Publication Data

Kennedy, Barbara A., 1943–
    Inventing the Earth : ideas on landscape development since 1740 / Barbara A. Kennedy.
        p. cm.
    Includes bibliographical references.
    ISBN-13: 978-1-4051-0187-5 (hard cover : alk. paper)
    ISBN-10: 1-4051-0187-3 (hard cover : alk. paper)
    ISBN-13: 978-1-4051-0188-2 (pbk. : alk. paper)
    ISBN-10: 1-4051-0188-1 (pbk. ; alk. paper)   1. Geomorphology–Philosophy–History.
2. Earth sciences–Philosophy–History.   I. Title.

    GB406.K46 2006
    551.41—dc22

                                                                        2005011628

A catalogue record for this title is available from the British Library.

Set in 10.5/12pt Dante
by Graphicraft Limited, Hong Kong
Printed by Replika in India

For further information on
Blackwell Publishing, visit our website:
www.blackwellpublishing.com

Contents

1-24-07 41.95

Illustrations

Preface

This is not a textbook on the History of Earth Science. It is a series of linked essays focusing on what I consider to be key episodes in the development of late twentieth century geomorphology in general and Anglo-American fluvial geomorphology in particular. There are also speculations about the ways in which key ideas may change as the twenty-first century opens.

The book derives from undergraduate lecture series given at the Universities of Manchester and Oxford, over the period 1973–1999 and from a variety of contributions over a 30 year period to the geographical and geomorphological literature focusing on historical figures and ideas. All of which stems, ultimately, from the graduate seminars run by Professor Arthur Strahler at Columbia University in the 1950s. Strahler's students Richard Chorley and Mark Melton pursued similar formats at Cambridge and at the University of British Columbia (UBC). In each case, the class of undergraduates or graduates was assigned reading of 'key' historically significant articles, for which they had to produce written abstracts. Debate then ensued concerning the content and continuing value of the material. The course I attended at Cambridge in the early 1960s was run jointly by Chorley and David Stoddart, the latter already deeply committed to a view of geography and geomorphology linked generally to the European Enlightenment tradition and particularly to the works of Charles Darwin. Chorley at that time was about to produce (in 1964, with Anthony Dunn and Robert Beckinsale) the first volume of The *History of the Study of Landforms*, whose impact was substantial. Probably without the influence of that work, this volume would not have been undertaken. The graduate Home Seminar at UBC, 1965–7, organized by Mark Melton, followed a similar format to Chorley and Stoddart's Cambridge course, but with a more overtly North American stance. An extremely important work introduced in the Melton seminar was the volume edited for the Geological Society of America by C.C. Albritton (1963) entitled *The Fabric of Geology* and most particularly G.G. Simpson's article on 'Historical science'.

Both sets of seminars appeared at the moment when Thomas Kuhn's notions of paradigms and scientific revolutions were freshly minted (1962).

The present work opens with a discussion of the way in which we arrive at 'scientific explanation'. In particular, it examines the use of T.C. Chamberlin's Method of Multiple Working Hypotheses (1890) as an essential tool for comparing competing explanations during the course of 'normal' science and its failure to assist in the evaluation of genuine new inventions (or incipient scientific revolutions): a key text and example is the American treatment of early plate tectonic evidence, as outlined by Naomi Oreskes (1999).

The next seven chapters deal more or less chronologically with the development of ideas which we currently view as axiomatic (or as part of our paradigm). These include (Chapter 2) the fundamental notion of 'deep time', first effectively proposed by James Hutton in the late eighteenth century, but not finally confirmed until the work of Ernest Rutherford, Arthur Holmes and Claire Patterson in the twentieth century. Charles Lyell's attempt to provide a 'scientific' basis for the study of the Earth (Chapter 3) required him to lean heavily on a notion of a net equilibrium of processes, in order to distance himself from all grand Theories of progressive change. This was undoubtedly a key factor in Lyell's general and persistent opposition to Louis Agassiz's dramatic, historically specific and distinctly non-equilibrium vision of a massive Ice Age (Chapter 4). This covering of land ice in the Northern Hemisphere was thought to have waxed and waned and, in the process, to have provided an alternative to the everlasting sea and its actions as a moulder of scenery. Charles Darwin (Chapter 5) employed Lyell's *Principles* as a textbook on the *Beagle* voyage and continued to accept Lyell's view of the dominance of marine processes for many years. But Darwin went well beyond Lyell in his acceptance of a variety of explanatory modes, including both Agassiz's land ice, and the significance of fluvial processes. This acceptance of glaciation and its myriad consequences in turn assisted acceptance of the key rôle of rain and rivers in the generality of earth surface sculpture, exemplified by the work of John Wesley Powell and Grove Karl Gilbert in the USA Southwest (Chapter 6). The 'normal' erosion by fluvial processes, plus a pseudo-Darwinian view of inevitable historical development underpinned the extraordinarily influential notion of the Geographical Cycle promulgated by William Morris Davis (Chapter 7). These historical ideas were explicitly rejected by the physically derived, reductionist views emanating from the work of Robert E. Horton (Chapter 8) and developed in rather different directions by Arthur N. Strahler and Luna B. Leopold, their colleagues and pupils.

The last chapter focuses on changes from the late twentieth century: in the attempts to incorporate the historical with the immanent ('abiding') processes of physics and chemistry; in the recognition of the ubiquity and importance of microorganisms in earth-forming processes; and in the continuing ambivalence of the view of *Homo sapiens* concerning the control or otherwise of natural forces.

The emphasis throughout is on selected individuals – heroes or villains – and their original work. Most illustrations are from original sources, or the author's photographs. There is some attempt to set cases in a broad framework of time and ideas, but no deep consideration of the history and sociology of science. The bias towards fluvial geomorphology necessarily omits other major figures, especially in the twentieth century.

A cast of principal characters and a glossary of key terms is provided as well as a general index. There is an extensive list of references, ranging from 1729 to 2005.

I am deeply indebted to the staff of the Rare Books Collections at the University of South Carolina at Columbia and the University of Cambridge, as well as the librarians at the Department of Geology, Cambridge and the School of Geography and the Environment, Oxford, for access to original material. My travels in pursuit of Charles Darwin were funded in part by the University of Oxford and in part by the Royal Society. Some of the illustrations were produced by Martin Barfoot and Ailsa Allen in the School of Geography and the Environment, Oxford.

My most profound debt of gratitude goes to Mrs Ria Audley-Miller, Deputy Academic Administrator of St Hugh's College, who has undertaken the production and reproduction of the manuscript; I should also like to thank the other members of the St Hugh's College Office who have rallied round at moments of crisis.

Professor Neil Roberts of the University of Plymouth suggested that my lecture course be turned into this book: I trust he will not be too disappointed.

Acknowledgements

I am grateful to three anonymous readers of the book's outline for their extremely positive support for the idea. Professor Michael Church, University of British Columbia, and Dr Nick Cox, University of Durham, read the full text of the original manuscript, with detailed – if conflicting – reactions. I have taken some of their key points on board, but I fear neither will be entirely satisfied by the finished product. I do hope, however, I have removed the factual errors to which they pointed: remaining errors are my own.

Finally, I am grateful to my geologist cousin, Dr Jana Hutt, for her general view on the manuscript. More especially, I should like to express profound thanks to Dr Adrianne Tooke, of Somerville College, who has read and re-read the text from the point of view of a general reader, to its immense benefit.

Barbara A. Kennedy
St Hugh's College, Oxford

Introduction

At the opening of the third millennium of the Christian era, it is interesting to take stock of the apparently incontrovertible truths that are held by at least Western scientists about the Earth we live on. They are, first and foremost, that this planet is almost unimaginably old (about 4600 million years, give or take); and its oldest rocks, dating to nearly 4000 million years, reveal a history of erosion, sedimentation, lithification and uplift (also known as the Geological Cycle) of almost equal antiquity. Third, we would see most of the surface of the planet as being in perpetual motion through the agency of plate tectonics; and, fourth, consider that the broad structures created by tectonism are continuously modified by a whole range of subaerial forces (wind, rain, rivers, the sea, glaciers and ice sheets). Finally, we think it indubitable that there has been endless variation in climatic conditions, such that different agencies have dominated in different global settings at different times.

Even 100 years ago, a large part of this 'received wisdom' was not available: there was no method of giving apparently absolute dates to rocks (although Ernest Rutherford had realized that the radioactivity discovered by the Curies provided a means to extend the planet's lifespan beyond the paltry 10 or 20 million years allowed by Lord Kelvin's vision of a continuously cooling spheroid – see Burchfield, 1990). There was recognition of widespread glaciation, but not of quasi-continuous shifts in temperature and concomitant fluctuation in land or sea levels, and there was no conception that the continents were other than fixed in place.

Go back a further 100 years and the accepted contemporary view of the age of the Earth may have been highly disputed, but many still would think in terms of thousands rather than million of years. The origin of rock strata as the product of the Geological Cycle rather than the Creation or Noah's Flood was a new and profoundly contentious concept. The notion of changes in the relative elevation of the land could be linked only to occasional earthquakes; and there was widespread disbelief that rivers, in particular, could achieve anything much except to shift the remaining debris of the Flood from place to place.

Back again, and the early eighteenth century saw – especially in England, France and Germany – struggles to reconcile the literal truth of the Biblical account of Creation (and Archbishop Ussher's chronology which dated that event precisely to 4004 BC) with the elegant rationalities of Newtonian mechanics. The varying nature of rock types, the problem of their fossil contents, the apparent rapidity of erosion by rivers when compared with the seemingly pristine character of dated, classical sites (see Davies, 1968) and the feeling that the Almighty had created all of the globe with human occupancy expressly in mind, led to theories (and Theories) which are in many ways at many removes from the thinking of the twenty-first century.

The object of the present volume is to try and show how our contemporary 'truths' have come to be accepted. It is not intended to maintain a Whiggish view in which the development of our current paradigms shows the ineluctable emergence of Truth from the morass of doubt and obfuscation. Rather, it is an old-fashioned attempt to chronicle how successive generations of natural philosophers and geologists and geomorphologists have come to invent the view of the Earth in general and of the Earth's surface in particular, which we currently accept as 'given'. There is absolutely no guarantee that such a vision will be seen as correct as the twenty-second century dawns. One aim of this account is to show the frailty of even the most impeccably scientific visions of the Earth, as the whole nature of beliefs and accepted wisdom shifts. (See Cohen, 1985.)

Since I am, by training, a fluvial geomorphologist the central viewpoint from which the story is told will be that of the changing significance of the action of rain and rivers. Since the work of Thomas Kuhn (1962) on scientific revolutions, sociologists of science have made us almost too aware of the importance of our specific temporal and geographical location in controlling what it is possible or academically acceptable to believe. Perhaps less attention has been paid – certainly in geomorphology – to the networks or families of academics by which received views of particular kinds are promoted and promulgated. I am undoubtedly the product of a very distinct scientific and geomorphological tradition: as are you . . .

Chapter 1
Inventing Scientific Explanations

'The distinctive aim of the scientific enterprise is to provide systematic and responsibly supported explanations' (E. Nagel, 1961, p. 15)

All the individuals that appear in this story would have considered themselves as rational, professional seekers after academic truth. But as Thomas Kuhn made us forcefully aware in his *Structure of Scientific Revolutions* (1962), the nature of scientific truth cannot be separated from the context in which it is sought. Whatever the variation in *nuance* given to the term, it is clear that Kuhn intends the concept of a *paradigm* to be understood as an overarching framework of beliefs and theories which determines the nature of the questions that it is permissible for scientists to ask and also the degree to which proposed explanations are seen as 'responsibly supported'. A mid-eighteenth century natural philosopher *could*, perhaps, have been a follower of the Roman philosopher and atomist Lucretius in his approach to physics – but in that case no-one would have thought his 'explanations' scientific. And he could not have proposed explanation in terms of quantum mechanics. If he wished to be a successful and accepted figure in the scientific world of his day, then he would necessarily have adopted the Newtonian paradigm.

It is the existence of paradigms which makes the scientific endeavour a collective – as well as a temporally and geographically specific – matter. Leonardo da Vinci, for all his undoubted genius, was a failure as a true scientist in that there was no body of opinion that could evaluate his observations and insights and make them the focus of new investigations.

For it is within the paradigm of the day that the overwhelming majority of scientific work takes place. Kuhn terms this 'normal science'. He also categorizes it as 'puzzle solving' and there is – as Imre Lakatos (1970, 1978) suggested – a strong tendency for normal science to pursue, selectively, those lines of puzzle solving which are thought of as 'fruitful' or as leading to full scale 'research programmes'. It follows, of course, that 'facts' as well as explanations are perceived as such only within the ambit of a particular paradigm.

However, Kuhn described the way in which 'normal science' can break down as more and more observations appear to conflict with tenets of the paradigm. This period (which Kuhn terms a crisis) may be long – as with the very gradual acceptance of the Earth's great age (see Chapter 2) – or relatively brief (the final acceptance of plate tectonics in the mid-twentieth century). One feature of the notion of scientific revolutions which has attracted much attention is that it seems to imply that the scientific community may choose a new paradigm without a fully elaborated test of all its consequences. As Kuhn states, 'All historically-significant theories have agreed with the facts, but only more or less' (1962, p. 146). Hutton and Playfair's vision of a fluvially sculpted landscape was demonstrably inaccurate as an explanatory tool for much of northwest Europe and yet it was quite widely adopted as a better paradigm than that of the Noachian Deluge (see Chapter 3). There has been enormous discussion about the nature of paradigm shifts and what they do or do not tell us about the distinctiveness of scientific 'explanation' (see for example Feyerabend (1975) and Lakatos (1978) for varying views). In particular, the adoption of a new paradigm seems to negate the basic rules of scientific testability, most particularly those set out by Karl Popper (1959) with his insistence not only that 'scientific' explanations must be testable but – crucially – that they should be *falsifiable*. If a new theory is patently at odds with at least some of the 'facts', how can it be a serious candidate for widespread adoption?

The present work shows that new paradigms in earth science have arisen, at least in part, through intuitive leaps on the part of their authors: they are, truly, inventions. Quite why one vision occurs as it does, to one individual in one time and place, is – in my view – impossible to pin down like a butterfly in a display case. Why did James Hutton and John Playfair, surrounded by the non-fluvial landscape of Scotland, see a fluvially dominated system of subaerial erosion? Why did William Morris Davis, who had worked in a meteorological observatory in Argentina and who corresponded freely with Albrecht Penck, the great co-author (with Brückner) of the concept of a fourfold Pleistocene glaciation, seemingly shrug off the notion of repeated, profound and recent climatic changes, in favour of the quintessentially fluvial concept of the Cycle of Erosion? (see Chapter 7). In both cases, it seems that it was the sense that the preferred theory explained, in some way, a whole range of often qualitative factors. There were those who clearly said 'Ah! So that's it . . .' and adopted the new ideas; and those who railed at their imperfect fit with reality. As Oreskes' masterly discussion (1999) of the difficulties preventing acceptance of the ideas of Continental Drift in the USA in the mid-twentieth century makes plain, the reality that you cannot apply Popperian standards to the evaluation of genuinely new visions (i.e. truly new paradigms) is very difficult to accept. She makes plain that, in this case the key problem for the American geological establishment was their overenthusiastic adherence to a brilliant device for evaluation of the relative reliability of explanations reached in the 'puzzle solving' course of 'normal' earth science: T.C. Chamberlin's 'Method of Multiple Working Hypotheses'.

Put forward in 1890, by a geologist (and bearing a close similarity to the earlier views – 1886 – of the great geomorphologist, Grove Karl Gilbert), Chamberlin's proto-Popperian approach to testing the validity of competing explanations has proved one of the most durable and sensible guides to the conduct of 'normal' historical science (i.e. that which works within a secure paradigm): it has been reprinted at least three times, most recently in a monograph aimed at ecologists (Hilborn and Mangel, 1997). By 'historical' science (see Simpson, 1963) I am referring to all those investigations which cannot be conducted according to the rules of experimental laboratory physics and where there is a significant (and often overwhelming) portion of any 'rationally supported' explanation which is contingent upon the precise sequence of events and/or the specific boundary conditions of one time and place. Astrophysicists and cosmologists are historical scientists alongside geologists, geomorphologists and ecologists. A key snag with all historical science is that we cannot re-run the natural experiments that have resulted in a river, a mountain or the solar system, although we may increasingly try to simulate them via computer models (see Chapter 9). As Douglas Adams made plain in *The Hitchhiker's Guide to the Galaxy* (1980), knowing that the answer to a question is '42' does not help to elucidate the nature of the original question. All historical scientists are faced with trying to explain configurations which are, in a real sense, the outcome of unsupervized natural experiments often of immense duration. Guessing what either the starting point was or the processes at work have been, is often very difficult.

There are those who would like to share Rutherford's withering scorn for the non-mathematical and the non-physical explanation, referring to biology as 'postage-stamp collecting' (quoted by Mayr, 1982, p. 33). Further, they will sometimes attempt to argue that to adopt historically contingent explanations is in some way to ignore the immanent or abiding (Simpson, 1963) regularities identified by physics and chemistry. This is a misconception and the career of G.K. Gilbert, in particular, shows how a great earth scientist combined the immanent and the contingent in variable and appropriate measure to meet the demands of satisfactory explanation (see Kennedy, 1992).

So how did Chamberlin propose that the historical scientist approach the formulation of responsibly supported explanations?

First and foremost, by the development of a whole family of hypothetical explanations for the observations under discussion. If possible, these should be set out before the investigation commences, so that it may be designed and conducted in such a way as to permit the evaluation of competing ideas. To take an example from my early work – on valley asymmetry (see Kennedy, 1976) – there are at least eight major competing theories which have been put forward to account for this phenomenon. By careful selection of field sites, it was hoped to eliminate all but two from consideration: the local erosional environment created by the stream in the valley bottom and the recent microclimatic regimes of the opposing valley sides (cf. Melton, 1960). (This indeed appeared to permit the development of explanations for the phenomenon.)

As Gilbert so graphically summed up this stage of an investigation:

'The great investigator is primarily and pre-eminently a man who is rich in hypotheses. In the plenitude of his wealth he can spare the weaklings without regret; and having many from which to select, his mind maintains a judicial attitude. The man who can produce but one, cherishes and champions that one as his own, and is blind to its faults. With such a man, the testing of alternative hypotheses is accomplished only through controversy. Critical observations are warped by prejudice, and the triumph of truth is delayed' (1886, p. 287).

The difference between a single, or Ruling Hypothesis (as Chamberlin terms it) and the multifarious family which both Gilbert and Chamberlin advocate, seems so profound and the advocate of the Ruling Hypothesis so blinkered that it may appear unlikely that any serious scientist would fall into that trap. However, not only are there all too many cases of precisely this myopia (and concomitant ill-temper) in the history of earth science, but there is a real problem when the rules of multiple hypothesis testing are applied to new paradigms; because a new paradigm has all the appearance of a Ruling Hypothesis. To quote Gilbert again:

'In the testing of hypotheses lies the prime difference between the investigator and the theorist. The one seeks diligently for the facts which may overthrow his tentative theory, the other closes his eyes to these and searches only for those which will sustain it' (1886, p. 286).

As Oreskes makes plain, the attempts to support Wegener's theory of Continental Drift by assisting the South African geologist Alexis du Toit to search for matching rocks, fossils and geomorphological phenomena in South America were deemed to smack too much of Ruling Theory, and there is a wonderful sequence of letters which shows that the investigation had to be cast in the form of 'testing' before funding could be assured (Oreskes, 1999, pp. 157–163). A similar, albeit smaller-scale conflict between 'theory' and Chamberlin's procedures for testing 'facts', was the truly scandalous treatment of J.H. Bretz by the American geological establishment for his temerity in proposing that the dramatic Channelled Scablands of the Pacific Northwest were the product of one or more cataclysmic glacial dam bursts (see Baker (1981) for a depressing account). In this instance, resistance to Bretz's ideas was – as with that to du Toit and Wegener – probably also in some measure due to the hint of the extraordinary in the mechanisms involved which, in turn, smacked of Creationist visions of Divine intervention in the details of earth sculpture (see later).

However, it remains the case that, for most earth scientists, for most of the time, Chamberlin's Method remains the best yardstick for evaluating the outcome of the puzzle-solving activities of their 'normal' scientific enquiries. The crucial operation is a genuine test of evidence against each competing hypothesis and not merely a rejection of those which do not fit, but a continuing (if, it is hoped, temporary) acceptance of those which cannot be eliminated. Where two or more hypotheses remain unfalsified, it may be possible to devise, or at least envisage some critical test or crucial piece of evidence. A good example comes from Charles Darwin's theory (1842) of the formation of coral atolls by subsidence of a central high island (in a contradistinction to Charles Lyell's (1830) hypothesis). Darwin concludes that the critical test of the difference between Lyell and himself would require 'a millionaire' to drill down through the centre of an atoll lagoon and see whether or not there was not merely an immense pile of coral, but, at great depth, a submerged mountain. The USA Government finally enacted the necessary drilling at Eniwetok Atoll, in 1952: Darwin's preferred hypothesis was supported.

It must be confessed that the explicit evaluation of conflicting explanations of phenomena has ceased to figure prominently in twenty-first century geomorphology, partly because of the shift to the observation of short-term processes. A brilliant example of the devastatingly effective use of the technique of multiple working hypotheses, before it was formalized and christened, is A.C. Ramsay's discussion (1862) of the probable glacial origin of lakes in the uplands of Europe and North America. The existence of these lakes – the Lake of Geneva was a key example – had been a severe stumbling block to the acceptance of Hutton and Playfair's fluvial views, despite the invention of the concept of the Ice Age by Louis Agassiz in 1837 (Carozzi, 1967: see Chapters 3 and 4). By a careful and extensive cataloguing of a range of five hypotheses which might account for the existence of the rock basins in which the lakes lie and successive elimination of all but glacial ice as a causal agency, Ramsay gives an elegant and effective example of Chamberlin's method as it should be applied.

In basic terms, of course, the Method of Multiple Working Hypotheses is no more than the form of evidence-testing which goes on in courts of law. The crucial thing is that it should not confuse the nature of the paradigm within which the particular piece of puzzle-solving is being conducted, with a Ruling Hypothesis which sees only one possible explanation for the phenomena under review.

One of the more thought-provoking discussions of the nature and limitations of (possibly) paradigms and certainly Ruling Hypotheses came, in 1926, from a most unlikely quarter. W.M. Davis not only invented the Geographical Cycle (Chapter 7) but promoted it quite ruthlessly and to the detriment of the careers of younger academics (see Chorley et al., 1964, 1973). In the course of his first visit to the West Coast of the USA, he let rip with a speech 'On outrageous geological hypotheses' which must have left his audience open-mouthed . . . Davis urged his hearers never to adhere too fixedly to any geological principle (by which he clearly envisages something akin to a paradigm) and, in so doing, has to be well ahead of his times in urging the mutability of 'received wisdom' . . . (see Kennedy, 1983).

However, in the heat of scientific endeavour, it is all too difficult to be sure whether any one study is merely part of Kuhn's 'puzzle-solving' phase and should, therefore adhere to the rules of falsifiability. If, of course, the work is opening up a revolutionary new vision – a true invention – then it cannot meet the full rigour of Popper's or Chamberlin's methods. When a truly new paradigm dawns, the principles of evidence change. Michael Ruse (1999) has recently examined the views of a group of evolutionists, starting with Erasmus Darwin. He concludes that the Popper/Chamberlin approach and that of Kuhn do, indeed fit different contexts. There is no single principle to be used in determining what constitutes a 'responsibly supported explanation'.

There is, however, one principle which all respectable earth scientists, at least since Hutton, have agreed upon: the Principle of Parsimony in Explanation known as Ockham's Razor. You may never invoke a supernatural cause to explain and you may not invoke an improbable cause, until you have demonstrated that all probable explanations fail under the test of Chamberlin's method. Or, as Conan Doyle has Sherlock Holmes put it in The Sign of Four (1890) 'when you have eliminated the impossible, whatever is left, however improbable, must be the truth.' This is the basic Principle of Uniformitarianism.

Uniformitarianism is, then, as Shea (1982) stresses, a rule for the conduct of science, *not* a great truth about the workings of nature.

It is most unfortunate, in my view, that the basic idea of the Principle of Uniformitarianism – which says nothing about the rate of operation of processes – became attached to the views of Charles Lyell (see Chapter 3), especially by Geikie, who dubbed him 'the great high priest of Uniformitarianism' (1905, p. 403). Lyell's views, which, as we shall see, encompassed more than a fair measure of floods, earthquakes and general cataclysms, somehow were taken to mean that earth processes always operated slowly and regularly. (A similar fate befell Darwin's views on evolution.) Thus various versions emerged of this mistaken view, one termed Gradualism and another Actualism (see Hooykaas, 1970). Any explanation which called for truly enormous forces (such as Bretz's Spokane floods) or even, as time went by, an event which seemed unusual on a human timescale, became know as Catastrophist or Neo-catastrophist. This is an extremely unfortunate and unhelpful development, in my view.

Hutton was setting out the proposition – as we shall see – that the Earth's structure and scenery could be explained adequately by invoking the range of geological and geomorphological forces currently to be seen in operation, always provided there was sufficient time. Lyell followed that lead. Neither wished nor felt one needed to invoke the Almighty's direct intervention to account for either the broad or the specific in the Earth's features. This Uniformitarianism is the most fundamental paradigm of modern earth science. Catastrophism, insofar as it exists in the twenty-first century, should be seen purely as an adjunct to Creationist views, a paradigm that is most definitely not held to be scientific by the earth science academy.

The influential seventeenth century natural philosopher, John Ray (1627–1705), worked under a paradigm in which the Earth itself was to be seen as a giant puzzle, specifically assembled and arranged by a Beneficent Being for the good of humankind. In that light, Ray's crucial questions about the features of the Earth were basically 'What has God intended them for?'. As Davies (1968, p. 112) points out, when Ray applied this reasoning to mountains, he came up with a range of answers which cannot but strike the modern reader as amusing: my personal favourite is that they are there to provide a suitable habitat for upland-dwelling animals . . . But we should beware: given the central paradigm of Ray's time and place, his question was considered scientific and his answers were 'responsibly supported'. Our own – we feel – impeccably scientific questions and answers may raise just the same smiles in 300 or 400 years time . . . or even sooner. Who can tell?

When our story of the invention of Earth in its modern guise opens, in the mid-eighteenth century, the key paradigms under which the increasingly professional class of natural philosophers in Europe were working were, roughly:

■ that the Bible, if not absolutely literally true, was, nevertheless of direct Divine authority;

- that, although Archbishop Ussher's chronology might be challenged, the Earth had quite patently been created, by the Almighty, at a period at best some tens of thousands of years ago (see Chapter 2);
- that the sole major change to the Earth's surface since the Creation had been the event of Noah's Flood, which had been directly Divinely engineered;
- that the Almighty, in his intervention, was of course conversant with the principles of Newtonian mechanics.

Within this framework, the fledgling earth scientists were wrestling with four interconnected problems.

1 What (and when) was the origin of the Earth?
2 What were the types of rocks and what were their origins? Did they all have the same origin?
3 What was the nature and origin of the fossils found in some rocks? If they appeared to be of animals and plants which no longer existed, how could that be reconciled with Noah's Divine mandate?
4 What was the nature and origin of the Earth's topography?

A crucial problem was that not all of these four questions could be answered in one fell swoop, although the proliferation of cosmological Theories through the seventeenth and eighteenth centuries all had, in varying degrees, a good try (see, for example, Davies, 1968; Gould, 1987). Burnet, Whiston, Woodward, Bourguet, Nogaret and even Buffon (1749) all tried to combine Divine control with natural forces largely adhering to Newtonian principles, so as to create the Earth, its rocks and their included fossils and its topography in either one go (the Creation) or, increasingly, two (the Creation and Noah's flood). An excellent example of the kind of thinking which prevailed comes from Louis Bourguet (1678–1742) in his *Traité des petrifications* published in 1742. This work is concerned mainly with trying to resolve the question of the origin of rocks; and 'the Author will try to show, by reasoning from the nature of the rocks themselves, that most of these stones are products either of the original fashioning of the Earth, or of its rejuvenation during the deluge' (1742, pp. iv–v; my translation). Further, he explicitly links the Divine and the modern, scientific: 'All fields of Physics are constantly producing new discoveries: and something which is of tremendous satisfaction to the Christian Philosopher is that all of these combine most admirably to confirm the factual truths upon which revealed Religion is founded' (p. xiv; my translation).

One key element in these discussions, which has been largely overlooked (but see Kennedy, 1997a), was the fascination with the identification of topographic regularities, both in the shape of the continents (Buffon, Article VI, 1749) and, more especially, in the non-random arrangement of mountain masses and volcanoes. These regularities became easier to spot with the development of atlases in a major way in the late seventeenth century and, in 1733, the production, by Doppelmayr in Nuremberg, of a genuinely modern map of the two hemispheres, based on astronomical observation (Bagrow, 1985). Although not all the Earth's surface had yet been charted, and there were continuing problems in precisely fixing longitude (cf. Sobel, 1995), this map and its successors provided the framework upon which the growing numbers of students of rocks and relief could fasten the plethora of topographic observations which the later eighteenth century produced. The voyages of Anson and of Cook, the travels of Linnaeus in Lapland, of Pallas in Russia and de Saussure in the Alps, the explorations of North America (cf. Jefferson, 1788) and, at the very end of the century, the monumental travels of Humboldt and Bonpland in Central and South America, all brought observations of new mountains, in particular, and their locations were eagerly noted and aligned with former ranges. The obsession with topographic geometrics persisted in France, in the person of Elie de Beaumont until the middle of the nineteenth century (cf. 1852); and James Dwight Dana in the USA betrays traces of a similar fascination (1869).

With the superiority of hindsight we can see that a fixation on the need to explain an increasingly complex global topography with the other three aims listed above was doomed to failure. Not that the observations were unscientific. It was, for example, widely noted that mountains generally occur in long ranges, not at random. Further, that mountain ranges near coasts nearly always have active volcanoes: indeed, that most active volcanoes are near the sea (cf. Cuvier, who asks 'Why are almost all active volcanoes close to the sea?' (1810, p. 190): and he was echoing J.R. Forster, 1778). Evidently, the entrance of sea water into the coastal rocks might be thought of as responsible for volcanic eruptions . . . Before you laugh, pityingly, you should realize that one of the latest ideas of plate tectonists is that the oceanic plates which are subducted beneath continental margins, pushing up mountain ranges such as the Andes *en route*, carry loads of very wet sediments and as this water is expelled at depth it does indeed fuel volcanoes.

But the key question which had to be answered, before the course of modern earth science could get underway, was that concerning the age and origin of the planet.

So that is where this story will begin.

Chapter 2

Inventing the Age (and Origin) of the Earth

'(I)f the succession of worlds is established in the system of nature, it is in vain to look for any thing higher in the origin of the earth. The result, therefore, of our present enquiry is, that we find no vestige of a beginning, – no prospect of an end' (Hutton, 1788, p. 304).

INTRODUCTION

At the time of writing, it is the general scientific view that the Earth is something under 4600 million years old and accreted from a mass of debris, which was originally cold and which has been heated from its core during the planet's life-time by energy released by radioactive decay. The rôle of radioactivity is, there-fore, seen as absolutely crucial: it not only permits us to place a date on the planetary origin (first achieved to the present precision as recently as 1953 by Claire Patterson), but it is also thought to be the principal source of endogenetic heat and energy (see Holmes, 1965, pp. 376–8; Bryson, 2003).

It is difficult to underestimate the importance of this basic, dual vision to all our scientific thoughts about both the Earth and its inhabitants. If the agreed timescale were to be either substantially reduced – or lengthened – then it would necessarily involve a massive reorganization of our thinking about the nature and rate of all geological and geomorphological and biological processes. Our current notion of the Earth's age forms, then, the most fundamental part of our contemporary paradigm in the earth sciences.

At the start of the eighteenth century, the ruling vision of the Earth's origin was that 'as told to Moses' in the first chapters of Genesis: of Divine creation, over a very brief period. Let me remind you of the wording, in the King James version of the Bible (Genesis, Chapters 1 and 2):

Ch. 1 v. 1   In the beginning God created the heaven and the earth.

        v. 2   And the earth was without form and void; and darkness was upon the face of the deep. And the spirit of God moved upon the face of the waters.

        v. 3   And God said, 'Let there be light and there was light . . .' and God divided the light from the darkness.

        v. 5   And the evening and morning were the first day.

        v. 7   And God said, 'Let there be a firmament in the midst of the waters and let it divide the waters from the waters . . .'

v. 8   And God called the firmament Heaven. And the evening and the morning were the second day.

v. 9   And God said, 'Let the waters under the heaven be gathered together into one place, and let the dry land appear . . .'

v. 10   And God called the dry land earth, and the gathering together of the waters called he the seas . . .

v. 11   And God said, 'Let the earth bring forth grass, the herb that yields seed . . . and the fruit tree . . .'

v. 13   And the evening and the morning were the third day.

Ch. 2   v. 2   And on the seventh day God ended his work which he had made; and he rested . . .

Archbishop James Ussher, in the mid-seventeenth century, undertook a monumental and scholarly investigation of the Biblical sources and concluded that the Creation had occurred in 4004 BC. Further refinement to this date was provided by Dr James Lightfoot, who stated that the Trinity's act of Creation was complete 'on the 26th October 4004 BC at 9 o'clock in the morning' (Lightfoot, 1654, quoted in Chorley *et al.*, 1964, p. 13).

Given this basic framework, the only further event of any significance was considered to have been Noah's Flood, which was held to have occurred in 2348 BC (Chorley *et al.*, 1964, p. 97). Genesis again (Chapters 7 and 8):

Ch. 7   v. 11   In the six hundredth year of Noah's life, in the second month, the seventeenth day of the month, the same day were all the fountains of the great deep broken up, and the windows of heaven were opened.

v. 12   And the rain was upon the earth forty days and forty nights . . .

v. 17   And the flood was forty days upon the earth . . .

v. 19   And the waters prevailed exceedingly upon the earth, and all the high hills, that were under the whole heaven, were covered.

v. 20   Fifteen cubits upward did the waters prevail and the mountains were covered . . .

v. 24   And the waters prevailed upon the earth an hundred and fifty days.

Ch. 8   v. 1   And God remembered Noah . . .

v. 5   And the waters decreased continually until the tenth month; in the fourth month, on the first day of the month, were the tops of the mountains seen . . .

v. 13   And it came to pass, in the six hundredth and first year . . .

v. 14   in the second month, on the seven and twentieth day of the month, was the earth dry.

Gordon Davies, in his splendid consideration of early ideas about the sculpture of the Earth's surface (1968), looks closely at the way in which the Earth's age and the appearance of erosional processes were variously combined, in the mid- and latter part of the seventeenth century according to the nature of the prevailing religious paradigm. When John Calvin's views were dominant, the 'Earth in decay' was seen as crumbling, under the onslaught of rain and torrents, at such speed that by the Second Coming (or Millennium, the year AD 2000) it would be totally destroyed and set for Divine reconstitution as a paradisiacal habitation for the Elect. When Calvin's doom and gloom was displaced by the more human-friendly Arminian doctrine, the shift in paradigm necessitated a reinterpretation of the 'facts' of terrestrial decay, giving rise to what Davies christened 'the Denudational Dilemma'. If there were only *c*. 6000 years of earth history and, if torrents and the sea were busily attacking and eroding the habitable portions of the globe, then how could this be squared with an active, Divine Providence? And, more difficult, how was it that many features described by the Classical authors, which could be quite securely dated, seemed often to have escaped Time's ravages?

One neat, if not altogether convincing solution was provided by Robert Hook(e) (1635–1703): 'the Earth itself doth, as it were, wash and smooth its own Face, and by degrees to remove all the Warts, Furrows, Wrinckles and Holes of her Skin, which Age and Distempers have produced' (quoted in Davies, 1968, p. 48).

Kuhn would describe the situation in earth science, by the early eighteenth century, as one of incipient crisis. It became clear that the date of Creation needed to be pushed back in time, whilst maintaining the existence of both a Creator and a moment of Creation. The first and most 'scientific' attempt to fix on a date by the most up-to-date principles of physics and the use of mathematics was made by the great French natural historian, Georges-Louis Leclerc, Comte de Buffon (1707–1788).

BUFFON'S APPROACH: VIA EXPERIMENTAL PHYSICS

Probably no other *savant* of the eighteenth century was as productive as Buffon, although he could devote only about half of each year to his researches (Roger, 1970), as every year, from 1739 until his death, he was based for six months in Paris as *Intendant* of the Jardin du Roi. His ideas about the Earth – both its origin and sculpture – evolved quite markedly between his *Premier discours* of 1749, which included '*Histoire et théorie de la terre – Preuves de la théorie*' and the 1775 and 1778 volumes of his *Histoire naturelle* – in particular, the 1778 work, which was subtitled '*Des époques de la nature*' (for a complete bibliography of Buffon, see Piveteau, 1954).

Buffon's 1749 ideas hinted that the Biblical account was less than satisfactory. This was especially true in the Second Discourse, on the History and Theory of the Earth, where the rôle of the Flood, in particular, was treated ambiguously even though Buffon seems to accept Ussher's chronology. Despite – or perhaps because of? – his rousing call to employ the joint powers of Mathematics and Physics, his conclusions were bitterly attacked by Theologians from the Paris Faculty and he was forced to publish a pretty grovelling retraction (dated 12 March 1751: in Book IV of his *Histoire Naturelle*, 1753)

'I declare:
First, that I had absolutely no intention of contradicting Holy Scripture; that I believe whole-heartedly in all that bears upon the Creation, whether it is the timeframe, or the sequence of events; and I renounce anything in my book, concerning the origin of the Earth and all that could be seen as contrary to the

Mosaic account, since I only put forward my ideas about the formation of planets as a purely philosophical suggestion.' (My translation.)

In that early work, he had made the very interesting suggestion that the saltiness of the sea might serve as a chronometer: as Lovelock shows (1979), it does not, but it would certainly have given a major extension to the Mosaic sequences.

However, by 1775, Buffon had conducted a series of meticulous experiments on his estates at Montbard, in Burgundy. These involved heating spheres of different composition and dimensions and then observing the time which each took until the family milkmaids could *'le toucher sans se brûler'* . . . (Buffon, 1775, edn Sonnini, L'an. IX, p. 79). One wonders quite what became of the cows in the meantime . . .

These experiments demonstrated to Buffon's satisfaction both that the Earth could not be considered as a simple sphere of iron and that the experiments could allow him to place dates on various key events, with a comparison to what might be expected in the case of Jupiter. These various estimates are set out in Table 2.1.

It followed, Buffon surmised (p. 267), that the globe which became cool to the touch *c.* 34,000 years from its origin, had been habitable for some 40,062 years and, in his view, would remain so until 168,123 from the origin. In the 1778 *Epoques*, he further considers that it would have taken 25,000 years before the Earth was cool enough for there to be rain (Sonnini edn, p. 252). It is interesting, as Roger (1970) reports, that Buffon in a manuscript note suggests that a period of 3 million years might be a better estimate for the age, judging from the evidence of sedimentation. So, by following a Baconian and Newtonian route, starting with an incandescent mass, Buffon could demonstrate rather convincingly that the Earth was at least almost twelve times the age ascribed to it by Ussher. Within this expanded framework, Buffon's earlier (1749) attempts to ascribe rocks and global topography to, first, the cooling processes (1749, Article VII) and, second, to the Noachian Flood (Article VIII) could be dramatically re-cast, and were, in the *Epoques de la nature* of the 1778 volume (page references to the Sonnini edition, Year VIII, vol. III).

Table 2.1   *Buffon's experimental estimates of times for planetary cooling, in years (data from Sonnini edition, pp. 78–263)*

	BECOMES SOLID TO THE CORE	COOL ENOUGH TO TOUCH	COOLED TO PRESENT TEMPERATURE	TOO COLD (1/25 OF PRESENT TEMPERATURE)
Earth, if entirely of iron	4026	46,991	100,696	?
Earth, if made of glass, marble, sandstone and limestone	2905	33,911	74,047	?
Earth as it is, approximately (iron core)	2936	34,270.5	74,832	168,123
Jupiter	9433	110,118	240,451	483,121

'There is absolutely no doubt that (Nature) today is very different from its condition in the beginning and also from what it became over successive time periods. It is these various alterations which we call Nature's epochs' (pp. 159–60; my translation).

He identifies six distinct epochs. These are:

1st Epoch	The Earth's substance is still hot and liquid, but developing its form of an oblate spheroid, by the effect of rotation.
2nd Epoch	Saw the consolidation of the Earth's materials and the creation of the great masses of glassy, vitrified and unfossiliferous rocks.
3rd Epoch	The sea extended over what are now the habitable lands and provided a home for the shelled creatures, whose remains have formed the calcareous rocks.
4th Epoch	Saw the retreat of those seas from what are now the continents.
5th Epoch	Saw elephants, hippopotamuses and other southern creatures living in northern lands.
6th Epoch	This, the latest period, saw the separation of the two great continents (of the Americas and Eurasia with Africa: see Figure 2.1).

And what about the evidence of the Bible? 'What must we understand by the six days which the holy scribe defines so exactly, and counts so precisely, if not six periods of time or six series of events?' (p. 201).

In conclusion, Buffon states, triumphantly, that he has reconciled for ever 'la science de la Nature avec celle de la théologie' (natural science with theology) (p. 208). But, alas, as Cuvier stated, these ideas 'came too late to have even a temporary success' (1810, p. 197; my translation). We shall return to some of Buffon's thoughts about Earth sculpture in Chapter 3, but we must now turn to a vision of the Earth's history which came from quite different premises.

JAMES HUTTON'S OBSERVATIONS

James Hutton (1726–1797) was a Scot who had studied chemistry and law, before qualifying as a doctor of medicine (see Dean 1992; McIntyre and McKirdy, 2001; Baxter, 2003; Repchek, 2003). However, unlike Buffon, Hutton in 1750 was able to retire to his estate in Berwickshire, where he undertook to become a 'scientific' farmer in the most up-to-date Norfolk fashion and also to pursue and enlarge his range of other 'philosophical' pursuits. His first love, chemistry, was developed by a series of joint experiments with Sir John Hall and after 1768 he became part of the key Edinburgh scientific circle of the Enlightenment. From chemistry, he was led to the scientific and philosophical study of mineralogy and, ultimately, to a view of geology, which he developed in a paper read to the Royal Society of Edinburgh in 1785. This was published, in 1788, in the first volume of their *Transactions*, as *Theory of the Earth; or an Investigation of the Laws observable in the Composition, Dissolution and Restoration of Land upon the Globe*. As John Playfair (1747–1819), the friend and champion of Hutton and his views, stated in his Life of Hutton:

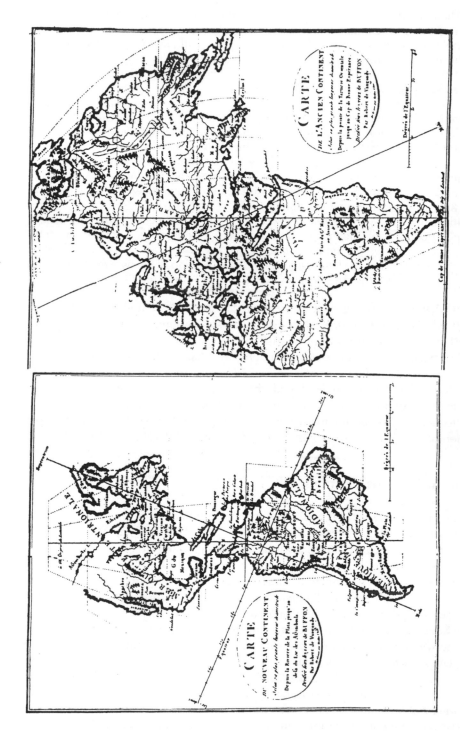

Figure 2.1 *Buffon's map to show the geometric regularities of the two continental land masses (1749).*

'The object of Dr HUTTON was not, like that of most other theorists, to explain the first origin of things. He was too well skilled in the rules of sound philosophy for such an attempt; and he accordingly confined his speculation to those changes which terrestrial bodies have undergone since the establishment of the present order, in as far as distinct marks of such changes are now to be discovered.

With this view, the first general fact which he has remarked is, that by far the greater part of the bodies which compose the exterior crust of our globe, bear the marks of being formed out of the materials of mineral or organized bodies, of more ancient date. The spoils or the wreck of an older world are everywhere visible in the present, and, though not found in every piece of rock, they are diffused so generally as to leave no doubt that the strata which now compose our continents are all formed out of strata more ancient than themselves' (Playfair, 1805, p. 52).

The key difference, then, between Buffon and Hutton is not a simple distinction between physics and chemistry as the preferred explanatory mode, but is far more profound. When Hutton looked at the evidence set out in scenes such as that shown in Figure 2.2 (from his 1795 two volume elaboration which was so convoluted in style in the presentation of his views that it obfuscated rather than clarified) he saw in his most telling phrase 'no vestige of a beginning, – no prospect of an end' (1788, p. 304).

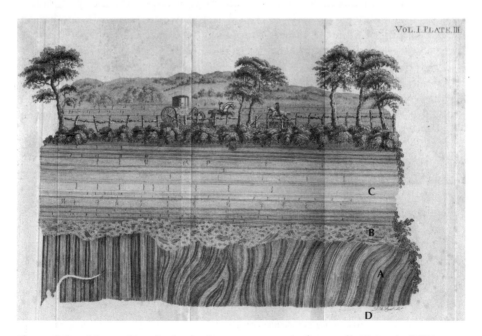

Figure 2.2 *'No vestige of a beginning, no prospect of an end'. Hutton's 1795 Jedburgh Unconformity. For key, see text.*

If we 'unwrap' the exposure of rocks near Jedburgh (in the Cheviot Hills of the Scottish Border), shown in Figure 2.2, what do we find?

At the base of the section, A, there are alternating layers of dark and light rocks (now identified as Silurian or *c.* 420 million years in age). These stand on end and also show arching and buckling. They are made up of material which, although much altered by heat and pressure, clearly was the product of the erosion of a land mass and the transportation of its remnants until laid down in a body of water. The striping shows that there were successive episodes when different kinds of deposit were created. As all waterlain deposits are effectively horizontal, there must have been some major upheaval and overturning of these beds, to produce their near vertical position. That upheaval seems to have brought the beds above sea level, since at B, there is evidence of what is termed an unconformity (that is, a more or less plane surface cut across the dipping rocks representing an extended period of subaerial denudation). Then, it seems, the area sank again, since the upper beds (C) were laid down horizontally (these are Devonian sandstones, *c.* 380 million years old) and lithified. Most recently of all, the whole lot has been uplifted again and the River Jed (D) is now busy cutting its way down through a slice of earth history.

When, to the interpretation of scenes such as that near Jedburgh, Hutton added his observations of the slow rate at which erosion and sedimentation occurred in the Scotland of his day, he was convinced that the time which must have elapsed since the moment of Creation, was effectively indefinite.

Unlike Buffon, Hutton did not have recourse to metaphorical interpretation of Genesis to support his Theory. A critical factor, which perhaps helps explain Hutton's freedom from the Mosaic chronology, was that he was a Deist and as such, considered that it was the Earth – not the Bible – which revealed the purpose and handiwork of God.

As with Buffon, however, Hutton's views were scarcely received with massive enthusiasm: it took the elegant glossing by John Playfair (1802) to rescue them from obscurity. We shall return to the broader implications of the Huttonian vision in Chapter 3.

EARLY NINETEENTH CENTURY REACTIONS

I have already quoted Cuvier's 1810 dismissal of Buffon's great theoretical work, *Epoques*, as coming '*trop tard*' (too late). In many respects, the same could be said of Hutton's *Theory*. For one thing, both Theories and *Théories* were seen as old-fashioned, dogma-ridden and, frankly unscientific. Nowhere can this be seen more clearly than in the evaluation made in 1810 by Georges Leopold Dagobert, Baron Cuvier (1769–1832). On behalf of the *Classe des Sciences physique et mathématique de l'Institut de France*, Cuvier presented to the Emperor Napoleon a report on the progress of the natural sciences since the Revolution of 1789 as well as a survey of their present condition. We have seen how Buffon's late *Théorie* was dismissed. Hutton rates only one mention (p. 188) concerning the nature of trap rock (injected volcanic material such as forms the Whin Sill of Northumberland) and even then his views on this specific matter are scarcely accurately or generously treated. Cuvier prefers to support the wide-ranging but explicitly parochially based views of the German mineralogist Abraham Gottlob Werner (1749–1817) which were derived from observations of the sequence of rocks found in Saxony, a sequence which Werner related to the differential precipitation of materials as the Flood subsided. As far as other theories went, Cuvier was distinctly damning with faint praise:

'How ever much genius, how ever much imagination was needed to invent these systems and make them fit the facts, yet we cannot include them in this picture of scientific progress: rather they have tended to obstruct the true path since they encouraged the belief that one did not need further observations . . .' (p. 198: my translation).

A similarly dismissive view of Hutton was to be found in Charles Lyell's survey of geology at the outset of the first volume of the *Principles of Geology* (1830), and the innocent reader would have had no way of knowing that Lyell, in fact, had become one of the chief promulgators of the Huttonian vision, most especially with respect to the acceptance of the idea of there being a vast expanse of geological time and the abandonment of any attempt to speculate as to the precise nature and origin of the Creation (see below).

Hutton's unorthodox treatment of both time and Genesis led to apoplectic rebuttals by scientific clergymen such as Richard Kirwan (whose 1799 *Geological Essays* also took repeated swipes at Buffon). But there were others – the Scot, John MacCullough, for example – who seem to have seen Hutton's views as so sensible that they did not really warrant discussion (cf. 1814, 1816). The French émigré César-Auguste Basset, who translated Playfair into French in 1815 recounts enthusiastically the fit between Hutton and Playfair's vision and the scenery of Great Britain. For the most part, the significance of Hutton's ideas about the Earth's enormous age tended to be overlooked in the battles over his associated views about both the origin of rocks and the origin of relief. In the former instance, Hutton was a Plutonist in that he saw the processes of lithification and uplift as directly linked to the heat of the Earth's core. (This undoubtedly stemmed from his joint experiments with Hall.) The Plutonic notion of the origin of both unfossiliferous rocks and of primary relief encompassed the igneous origin of granite and the explicit linking of intrusive sills (such as the Whin Sill) to modern volcanic lava extrusions. Werner in contrast, held for a watery origin of all rocks except for the emissions of modern, active volcanoes: his Neptunian ideas were founded on premises closely akin to Buffon's Second Epoch, in that the hard, glassy rocks such as granite, without fossils, must be the first and oldest precipitates from the Flood.

One of Werner's keenest supporters was Robert Jameson (1774–1854) who became Professor of Mineralogy at Edinburgh (and, when he lectured to the young Charles Darwin, bid fair to put him off geology for life . . .). His three volume *System of Mineralogy* (1804–8) included, as Part III of Volume III, a discussion of the *Elements of Geognosy* (1808) in which his views on the inadequacies of Hutton (and Playfair) are expressed at some length and with some vigour, in the third chapter. After a fulsome recognition of 'the comprehensive mind of WERNER' (p. 41) he goes on to excoriate.

> 'those *monstrosities* known under the name of *Theories of the Earth*. Almost all the compositions of this kind are idle speculation, contrived in the closet, and having no kind of resemblance to any thing in nature. Armed with all the *facts* and inferences contained in these visionary fabrics, what account would we be able to give of the mineralogy of a country, if required of us, or of the general relation of the great masses of which the globe is composed? Place one of these speculators in such a situation, and he cannot give a rational or satisfactory account of a single mountain' (1808, p. 42).

This criticism shows, plainly, that Hutton's *Theory* was to be dismissed on multiple grounds. It did not address one of the key elements of early earth science speculation, namely the broad structure of the continents. To those of a determinedly eighteenth century cast of mind you had to explain the origin of the Planet (Hutton failed to do that, too) AND the nature of its rocks (he did, but in a way in opposition to Werner's widely respected ideas). Jameson states firmly that he had visited the key Huttonian sites where veins of granite were described as intersecting other formations (so the granite must be younger than the rocks it was intruded into). 'I have convinced myself,' (says Jameson, 1808, p. 110) 'after a careful examination of the places . . . mentioned, that they do not afford a single instance of granite veins shooting from old granite into the super incumbent rocks.' The generations of geology students who have made forced pilgrimages to Hutton's key sites must wonder what it was that Jameson did not see . . .

Finally, in his contemporaries' eyes, Hutton notably failed to consider, let alone explain, the regularities of mountain ranges, nor why the continents tapered at their southern extremities (obviously that was the product of the Flood draining southwards while the strata were still soft enough to mould easily).

Even Lyell, while crediting Hutton as the first to separate geology from cosmogony (1830, p. 4) includes his Theory in a list that runs from Werner to De Luc (a dogged upholder of non-Huttonian views). Lyell (1830, p. 88) further deprecates what he (falsely) holds to be the Huttonian 'doctrine of the sudden elevation of whole continents by paroxysmal eruptions'. Lyell considers it important to maintain Playfair's version of the Principle of Uniformitarianism (which he uses as the Frontispiece quotation in his 1830 Volume I):

> 'Amid all the revolutions of the globe the economy of nature has been uniform, and her laws are the only things that have resisted the general movement. The rivers and the rocks, the seas and the continents have been changed in all their parts; but the laws which direct those changes, and the rules to which they are subject, have remained invariably the same' (Playfair, 1802, p. 374).

Now Playfair, please note, says nothing about uniformity of *rates* nor of the temporally uniform distribution of land and sea, uplands and lowlands – both things upon which Lyell insists. However, having stated 'It was contrary to analogy to suppose, that Nature had been at any former epoch parsimonious of time and prodigal of violence' (1830, p. 88) Lyell goes blithely on to predict that the Great Lakes will at some point overflow and 'a deluge will lay waste a considerable part of the American continent . . .' (1830, p. 89). Nevertheless, for all his backing and filling about Huttonian merits, Lyell *is* categorical that both the evidence of sedimentation and that of the fossil record positively require the vast amounts of time which Hutton assumed.

So, nearly 50 years after Hutton's proposals, there was a major scientific author whose ideas were firm about 'the lapse of time'.

However, things did not develop in a straightforward fashion.

THE LATTER NINETEENTH CENTURY: CHARLES DARWIN VERSUS LORD KELVIN

In the first volume of his elaboration of his Theory, Hutton had stated 'Time, which measures every thing in our idea . . . is to nature endless and as nothing' (1795, p. 15). The term 'endless' was to prove a rod with which to beat those who held that the extent of geological time was indefinite: they could be accused of assuming it was infinite . . .

Nowhere was this misunderstanding more forcefully employed than in the intervention by the physicist William Thomson, Lord Kelvin (1824–1907) into the matter of the Earth's age. This was brought on by the passage in Charles Darwin's *Origin of Species* (1859, pp. 285–7), where Darwin attempts to demonstrate that the erosion of the topography of the Weald of southeast England would have required some 300 million years (the present estimate is a mere 60 million: see D.K.C. Jones, 1981). The story of Kelvin's attempts to demonstrate the absurdity and the *unscientific* nature of Darwin's speculation (since it did not derive from the first principles of physics) has been most engagingly recounted by Burchfield (1990) and will not be rehearsed in detail here. However, Burchfield's account does tend to leave the reader with the sense that Kelvin's impeccable reasoning (which started, like Buffon's, from the 'known' of an initial incandescent ball of matter and then proceeded by the best physics available), which produced an Earth at most 100 million years old and (on a bad day) a mere 5 million, was somehow more meritorious than Darwin's historical extrapolation . . . One is left with a strong feeling of agreement with the great German/American biologist, Ernst Mayr:

'The history of science has too often been written by physicists, who have never gotten over the parochial view that anything which isn't physics, isn't science' (1982, p. 14).

Be that as it may, as Burchfield explains, the discovery of radioactivity was used by Ernest Rutherford in 1904 to demonstrate that the 'known' first principles from which Kelvin had worked would have to be seriously altered. Burchfield recounts Rutherford's horror at finding himself about to contradict Kelvin's views at a public lecture where the very elderly Kelvin was apparently dozing in the audience. Rutherford saved the day by the statement that Kelvin had, presciently, maintained that his calculations would hold 'provided no new source of heat was discovered' . . . and radium was that source. Rutherford recounts 'Behold! The old boy beamed upon me . . .' (Burchfield, 1990, p. 164).

One cannot underestimate the significance of the involvement of the community of physicists with the problem of dating the Earth. Even after Rutherford's 1904 presentation, it took 50 years before Patterson established the presently accepted age of 4550 million years. (Although Holmes had arrived at a figure in excess of 2000 million as early as 1913: Church, personal communication.) But the clash between the methodology and prestige of the historical and the physical sciences reflected in the 'Darwin versus Kelvin' episode was to surface again, in the twentieth century, over Alfred Wegener's Continental Drift theory, which was deemed to be physically impossible (see Oreskes, 1999).

Hutton's two key contributions to the invention of the Earth we live on were: the rejection of the idea that the date of Creation could be simply read in the rocks; and his view that the observed pace of geological and geomorphological phenomena argued for a terrestrial time span far in excess of anything which could be reconciled with the Mosaic account. In a way, of course, the use of radioactive dating techniques does now let us read the evidence for Earth's origin in its oldest rocks . . .

But merely developing an adequate timeframe for the production of rocks and relief was, as I have already hinted, not sufficient. It was essential to invent a 'modern' view of Earth's surface processes. Step forward Sir Charles Lyell (1797–1875).

Chapter 3

Inventing 'Modern Earth Science': Charles Lyell and 'The Principles of Geology'

'[W]ater appears as the most active enemy of hard and solid bodies; and, in every state from transparent vapour to solid ice, from the smallest rill to the greatest river, it attacks whatever has emerged from above the level of the sea and labours incessantly to restore it to the deep. The parts loosened and disengaged by the chemical agents, are carried down by the rains, and, in their descent, rub and grind the superficies of other bodies' (Playfair, 1802, *Illustrations of the Huttonian Theory of the Earth*, SS95, p. 99).

INTRODUCTION

When the first of the three volumes of Charles Lyell's *Principles of Geology* appeared in 1830, it bore as its subtitle 'Being an attempt to explain the former changes of the earth's surface by reference to causes now in operation'. In this statement, Lyell was moving Hutton's invention of the ceaselessly operating Geological Cycle (uplift, subaerial denudation, deposition, lithification, renewed uplift, all of these more or less demonstrably still at work in the world of 1785: recall Figure 2.2) to the centre stage for scientific study of the processes responsible for Earth's rocks and relief. Hutton had firmly broken the link between the Creation and Noah's Flood and the origin of both strata and scenery. As noted in Chapter 2, it was these geological and topographical aspects of Hutton's ideas which aroused contemporary opposition, quite as much as his invention of Geological Time. Lyell was endeavouring to convince the growing community of earth scientists that 'causes now in operation' – running water, marine action, tectonic forces – were sufficient to account for the surface features of the planet. As we shall see, his attempts were not entirely satisfactory.

But before we turn to Lyell's *Principles*, it is useful to see what the previous century had produced in terms of observations which, to some degree, paved the way for Lyell. There are three areas which are interesting to examine. First, discussions about the origin of rocks (and especially the relation between modern lavas and igneous intrusions and extrusions such as the Whin Sill or the basalt of the Giant's Causeway); second, the widespread discussions of the broad sweep of global topography – continental shapes, mountain ranges and major valleys as well as local features of relief; and, finally, the evidence adduced for changing relative levels of land and sea.

PROBLEMS WITH HUTTON'S THEORY

THE ORIGIN OF ROCKS

Modern earth science recognizes three categories of rocks:

1   Igneous rocks are products of the Earth's molten mantle. They are generally vitrified and always unfossiliferous. They may be intrusive (particularly granite); or extrusive, having flowed over the surface as volcanic lavas. Some intrusions – such as the Whin Sill – were associated with now extinct volcanoes and can be thought of as lava which did not quite make it to the surface.
2   Sedimentary rocks, in contrast, are the products of the chemical and mechanical breakdown of other rocks of any type. They are laid down in horizontal beds on land or in lakes or, more generally, in the sea and, over time, are lithified by a complex combination of physical and chemical processes. Unconsolidated surface layers of detritus may be termed alluvium (laid down by rivers), colluvium or scree (produced by slope processes), sand and shingle (coastal sites), or till and drift (glacial products).
3   Metamorphic rocks are those – usually sediments – which have been subjected to intense heat and pressure and undergone both physical and chemical changes in the process. The injection of a granite mass into a pile of sedimentary strata will be accompanied by the creation of a 'metamorphic aureole', which is often the site of rich mineral deposits. It is possible to find fossils in some metamorphosed beds.

It is common to consider that, by 1800, there were two, contrasting views of rocks: that of Werner (see Chorley *et al.*, 1964, chapter 3), who saw the Earth's origin as a fluid ball, from which successive phases of chemical and mechanical precipitation in water created, first, the Primitive, unfossiliferous rocks such as granite and some slate; then the Transitional materials, such as other slates and greywackes, with fossils; then the Flötz (limestones, coals, basalts); Derivatives, such as sand and clay; and finally, the volcanic rocks. (Werner's categories do not correspond to the modern: limestone, for instance, is a sedimentary rock and basalt an igneous one). This *Neptunian* vision – relying on aqueous action – was challenged by Hutton's *Plutonic* system, where rocks were seen as continuing products of different combinations of Plutonic internal heat and pressure and subaerial denudation. Intrusive, non-fossiliferous igneous rocks such as granite could be younger than fossiliferous sedimentary rocks into which they were injected. Basalts were seen as genetically linked to volcanoes – past or present. The surficial sands and gravels covering much of Scotland were seen as products of fluvial action.

It is a mistake, however, to see these two views as the only ones available. A whole range of variants was in existence.

Buffon, from his earliest writings on the subject in 1749, saw that there was a clear distinction between the 'plutonic' core of granite, etc., created from the incandescent origin of the planet, and subsequent, fossiliferous and sedimentary layers, produced by the breakdown of former rocks. This view is summarized in his 1778 work as follows. 'The oldest and deepest layers are those of the primitive rocks of the globe, formed by drying the youngest and the nearest the surface have arisen from material transported and deposited as sediments by the movement of water' (1798, Sonnini edn, p. 103: my translation).

At the same time (1777–8), the German, Pyotr Simon Pallas (1741–1811), was describing the nature of rocks as deduced from the first of his extensive Journeys in Russia. The widely travelled Pallas criticizes Buffon roundly (p. 22) for his 'parochial' geological views. He, Pallas, speaks from a broader experience . . . In Pallas' vision, it is axiomatic not just that granite is the oldest rock, but that it is found *everywhere* in the core of mountains and 'nothing seems so probable as that we should consider this rock as the chief component of our world's interior' (Pallas, 1777–8, p. 25: my translation). But he thinks that the granite was not formed by heat, since heat will alter its nature. Out from the granite core, forming the highest peaks, as in the Caucasus, are 'the zones of schistose rocks which are always found on the flanks of the great ranges, together with the adjacent uplands of secondary and tertiary rocks' (1777–8, p. 28: my translation). Some of Pallas's most genuinely interesting observations concern what we would term the metamorphic aureoles, which he links (1777–8, p. 43, note e) to the former existence of volcanoes.

Not surprisingly, the significance of former volcanoes and their connection to basalts and other igneous rocks was first and best appreciated in Italy. Although Buffon (1778, p. 342) was also clear that many regions of France showed evidence for extinct volcanoes, one of the first to link them firmly to basalts was the Italian, Scipione Breislak (1750–1826). Several plates from the *Atlas* accompanying his *Institutions géologiques* (1818) are based on the earlier work (1778) of Barthélemy Faujas de Saint Fond (1741–1819) on the extinct volcanoes of the Vivarais and Velay areas of France. The latter, although impressed by the '*grandes et superbes chaussées formées par un assemblage de colonnes prismatiques*' (a broad and impressive highway created by a group of prismatic columns) bordering the rivers (Faujas de Saint Fond, 1778, p. 264), curiously wavers in his attribution of basalts to Plutonic, rather than Neptunian forces. A more idyllic scene, showing the competing rôle of volcano and river, appears as the Frontispiece to the present work. For Breislak, in his 1798 *Topografia fisica della Campania* had argued forcefully for the links between modern volcanoes and the characteristic layers of basaltic columns, described by Faujas de Saint Fond: looking at the illustration on the cover of this work it seems extraordinary that the connection could be doubted. Yet as late as 1811, Breislak in his text *Introducione alla Geologia* (translated into French, 1812 as *Introduction à la géologie*) felt it necessary to devote the whole of his last chapter to a consideration of basalt (1812, pp. 509–579). His clinching evidence for the Plutonic origin of basalts is that there may be as many as 20 layers intercalated with patently sedimentary strata: 'I do not see how the Neptunians can escape from the problem of finding a causal mechanism for ten, for fifteen, for twenty different precipitates, in one and the same part of the sea' (1812, p. 576: my translation).

But the ideas of Neptunists died hard.

They had more reason and more observational support when it came to the consideration of the masses of poorly consolidated and often highly mixed sands, gravels and boulders found at or near the surface over large stretches of north and west Europe (and also, of course, northeastern North America). Faujas de Saint Fond had commented on the occurrence of great beds of rounded stones and boulders separating some lava flows (1778, p. 281): they clearly related to some vast rush of water.

Some very pertinent questions were being asked about these and similar masses of debris. The Swiss, Horace Bénédict de Saussure (1740–99) in the last of his great four-volume *Voyages dans les Alpes* (1779–1796), asks:

'If a valley includes foreign rocks – that is, ones which do not originate from the adjacent mountains, then check how high they are found up the flanks of the mountains and ask: what is their origin and where can they have come from?' (1796, p. 495: my translation).

The indefatigable opponent of all Huttonian ideas, Jean André de Luc (1727–1817), a Swiss geologist long settled in England, in his great list of questions unanswered in the Huttonian system, especially by Playfair (1802, p. 103 and 352) states:

'Head II
The difference between the GRAVELS on the Heights and on the Declivities of the opposite sides of VALLIES, and that between those GRAVELS and the STONY STRATA which border them, are also in Contradiction to this Hypothesis' (1810, p. 11) (Valleys carved out of a block of terrain, by running water, should only contain examples of the rocks within each valley in their alluvium.)

Pallas's *Travels through the Southern Provinces of the Russian Empire in the Years 1793 and 1794* (translated into English, 1802–3) gave the usually favoured cause of such deposits:

'The vestiges of a former deluge are discoverable in the uppermost sandy and loamy strata, which are frequently intermingled with cylindrical stones; and, in the deeper clayey layers, there are likewise found the remains of marine productions . . .' (1802, p. 47). (The Flood had, of course drowned all topography, so 'foreign' stones could be expected in any valley.)

This ascription to Noah's Flood reached its apogee in the *magnum opus* of the great English eccentric and enthusiast, William Buckland (1784–1856) *Reliquiae Diluvianae or Observations on the Organic Remains contained in Caves, Fissures and Diluvial Gravel, and in other Geological Phenemena, attesting the action of an Universal Deluge* (1st edn, 1823).

From the outset, Buckland is perfectly clear about the source of the material which was so embarrassing to the Huttonian account:

'[T]he word *diluvium* . . . I apply . . . to those extensive and general deposits of superficial loam and gravel, which appear to have been produced by the last great convulsion that has affected our planet; and that with regard to the indications afforded by geology of such a convulsion, I entirely coincide with the views of M. Cuvier, in considering them as bearing undeniable evidence of a recent and transient inundation. On these grounds I have felt myself fully justified in applying the epithet *diluvial*, to the results of this great convulsion . . .' (2nd edn, 1824, p. 2).

Buckland devoted great observational power to demonstrating the discrepancies between the rocks of diluvial deposits and those simply being shed by weathering from modern hillsides (1824, p. 189); and also to the non-local origin of many of the pebbles found in the river systems of north Oxfordshire and south Warwickshire.

As we shall see (in Chapter 4) Buckland was to change his mind about the rôle of the Flood and, instead, embrace what we would feel is the correct ascription of the 'diluvium', namely to the work of land ice.

But, one way and another, Hutton and Playfair's *'théorie lumineuse'* (Breislak, 1812, p. 115) was not judged to fit the observed facts of surface geology. Although the Wernerians were most opposed to Hutton's concept of the igneous origins of strata, the most embarrassing mismatch between Theory and Fact was undoubtedly that between an essentially fluvial vision and the great piles of non-local diluvium.

There were further problems with the origins of topography.

CONTINENTS, MOUNTAINS AND VALLEYS

Our modern view is that the present continents are the quite recent (and essentially ephemeral) productions of the processes of plate tectonics. Those are thought to be driven by slow convection currents within the Earth's mantle. Mountain chains – such as the Andes – form where two plates clash together and one is forced to subduct down under the other. Earthquakes and volcanic processes assist in the elevation of the land. Although there are some very extensive areas of extremely ancient rocks (for example, in northern Canada: see van der Lee, 2001) these old 'shield' zones tend to be of low relief. The higher and steeper the mountains, the younger they are. Some of the largest valleys on the planet – the great rift valleys and intermontane trenches – are formed by plate tectonic processes. We think that the majority of valleys, however, are carved from the tectonically uplifted blocks, by running water and ice (see Kennedy, 1997a).

Hutton and Playfair effectively had nothing to say about the grand sweep of continental geometry or mountain ranges: they started with uplifted blocks and simply let denudational processes do the rest.

But in this, too, Huttonian ideas failed to address yet another area of contemporary concern, namely the tremendous excitement and theoretical effort which had been directed towards the grand lineaments of Earth's topography.

We may take Buffon as a key exemplar. In his 1749 *Histoire*, the VIth Article is entitled *Géographie* and is illustrated by a specially drawn and very carefully oriented map of the two great landmasses of the globe (Figure 2.1). The orientation of each land mass is '*selon sa plus grande longueur diamétrale*' (1749, p. 228) and the figure is reproduced in the 1798 Sonnini edition of Buffon's works, essentially unchanged. In the lengthy discussion, Buffon points out that this north–south banding of land and sea differs from the horizontal zones of Jupiter (p. 204). He notes that the longest diameter of the two land masses is closely similar (pp. 206–7) and 'We can add two rather remarkable facts to these observations; the old and the new world are virtually opposites in form; the old extends further north than south of the equator, whilst by contrast, the new lies more to the south than the north of the equator' (1749, pp. 208–9: my translation). The centre of gravity of the two masses – the old world at 16–18°N and the new at 16–18°S – seems to suggest that they are counterbalanced. And, finally, both continents are formed of two separate land masses, connected by very narrow isthmuses (p. 209).

The long diameters are thought to be significant because 'we can note that the oldest parts of the continents are those closest to those axes and are also the highest, whilst the newer portions are necessarily further away and also much lower' (pp. 209–10: my translation). In a stirring flight of fancy, he locates the great empires of America in the old highlands, whilst '*les sauvages au contraire se sont trouvé dans les contrées les plus basses et les plus nouvelles . . .*'

What was the significance of all this? Clearly – recall his First, Second, Third and Fourth Epochs (Chapter 2) – when the Earth's surface was still hot and plastic, it was deformable by rotation. At a later stage (the Fourth Epoch) the grand lineaments of the continents were created by the draining away of the sea, whilst influenced by the centrifugal forces of rotation. The assumption that mountains are both oldest and highest (the reverse of our current thinking) is linked, in my view, to a belief that there has been only one period of major topographic sculpture: the flat, 'young' lowlands are basically 'made ground' by those depositional processes which continue.

This thinking not only colours his views on global lineaments, but is applied to one of the earliest discussions of regional topography (1749, pp. 253–5; and Figure 3.1). He takes the landscape of north Burgundy, the Plateau de Langres, as an exemplar of the way in which the network of major 'mountains' and valleys must have been formed by the drainage of ocean waters. A key piece of evidence is held to be the earlier observation, by Louis de Bourguet (1729, 1742) of '*les angles saillans et rentrans*' or what we would term 'interlocking spurs'. The significance to Buffon's mind, was that the existence of the same rocks on each side of the valley showed that its sinuosities had been actively carved by moving water. (This also explained why, by and large, the two sides of a valley are of similar heights.) These '*angles saillans et rentrans*' figure very prominently in discussions of topographic origins, until Saussure observes that they are generally absent in major Alpine valleys, which have alternate broad and narrow stretches, instead (1779, Vol. 1; reprinted, 1803, p. 426): we consider that those truncated spurs and rock basins are the work of valley glaciers (see Chapter 4).

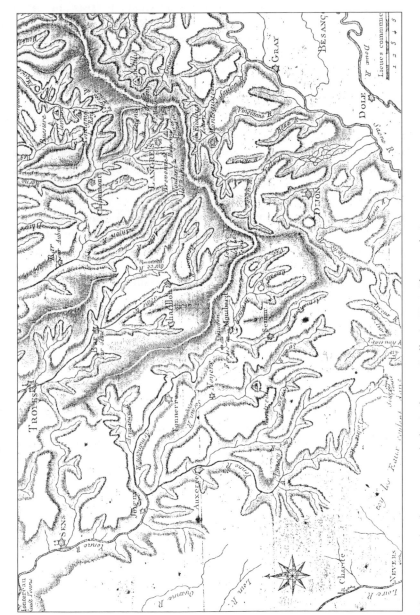

Figure 3.1 *The first geomorphological map: Buffon's 1749 Plateau de Langres.*

It is in the struggle to separate out the various scales of global relief that the clash of vision between the Huttonians and the rest was perhaps most profound, yet this is generally overlooked in modern discussions.

The preoccupation with geometric regularities, which persists through the accounts of the travels of Alexander von Humboldt (1769–1859) and his geological treatise (1823) to Elie de Beaumont's massive three-volume, 1852 work '*Notice sur les systèmes de montagnes*', can still be seen as late as 1870, in Noguès' discussion of applied geology.

When Johann Reinhold Forster describes the nature of the coast of 'New Holland' (Tasmania) as visited by James Cook in his second voyage, he goes to great lengths to assure his readers (1778, pp. 11–12) of the strong similarities between all the 'southern terminations' of the continents. It mattered.

It also mattered, as Robert Jameson points out, since the authors of 'those *monstrosities* known under the name of Theories of the Earth' (i.e. Hutton and Playfair) would be unable to 'give a rational or satisfactory account of a single mountain' (1808, p. 42).

We would now see these geometric problems as having three, separate solutions. We invoke plate tectonics to account for continental topography and mountain chains and rift valleys; we consider glaciation is responsible for many features of upland mountains and their valley systems; and then we endorse Playfair's emphasis on past and present fluvial action to account for probably the great majority of valleys.

But there was a final area where Huttonian ideas were thought to be at odds with both theory and observation.

CHANGING RELATIVE LEVELS OF LAND AND SEA

At present, we consider that sea level, relative to the land, can change by two mechanisms. One, isostasy, relates to changing loading of the crust – by ice or sediment or mountain building – and produces both up and down movements. The north of Scotland, for example, is still exhibiting isostatic rebound some 14,000 years after the major retreat of ice sheets whose weight depressed the surface: southeast England, in contrast, is sinking slightly as a counter-movement. The second mechanism is eustasy, which refers to changes in the volume of the seas. This can be influenced by tectonic forces and plate movements, but it is also linked to the amount of water locked up in ice sheets and glaciers. At the height of the most recent glacial period, world sea level appears to have been about 100 m lower than it is now. The simultaneous and often conflicting operation of isostatic and eustatic processes makes tracking the history of land-/sea-level changes extremely complex.

Hutton's Geological Cycle focused upon the denudation of a land mass. He inferred that the material eroded was deposited in the sea (Figure 2.2), altered by heat and compression and then at some point uplifted above sea level so that subaerial denudation would attack it anew. He was reasonably certain that there was an important connection between volcanic activity and the development of very high mountains (1795, Vol. I, pp. 146–7), but was forced to conclude 'We only know, that the land is raised by a power which has for principle subterranean heat; but, how that land is preserved in its elevated station, is a subject in which we have not even the means to form conjecture' (1795, Vol. I, p. 164).

However, he then goes on to consider the replacement of an old continent by a new one, saying 'The formation of the present earth' [i.e. the existing continents] 'necessarily involves the destruction of continents in the ancient world' (1795, Vol. I, p. 181). Further, 'the destruction of one continent is not brought about without the renovation of the earth in the production of another' (p. 183). (We would say that he considered that there was an equilibrium maintained between the amount of land and sea.) But, as we read on, this picture becomes more confused and the statements below are in flat contradiction to that just quoted:

'It is not necessary that the present land should be worn away and wasted, exactly in proportion as new land shall appear; or, conversely, that an equal proportion of new land should always be produced as the old is made to disappear. It is only required' (by his vision of Divine Providence) 'that at all times, there should be a just proportion of land and water upon the surface of the globe, for the purpose of a habitable world.

Neither is it required in the actual system of this earth, that every part of the land should be dissolved in its structure, and worn away by attrition, so as to be floated in the sea. Parts of the land may often sink in a body below the level of the sea, and parts again may be restored, without waiting for the general circulation of land and water, which proceeds with all the certainty of nature, but which advances with an imperceptible progression' (1795, Vol. I, pp. 195–6).

Just to further confuse the issue, he adds:

'In thus accomplishing a certain end, we are not to limit nature with the uniformity of an equable progression, although it be necessary in our computations to proceed upon equalities' (p. 197).

By this last, I think, he just means that we have to try to work out average rates, although reality will exhibit fluctuations from those averages.

I stated earlier (p. 17) that Hutton's 1795 treatise did little to clarify the thrust of his thinking. His friend and apologist, John Playfair (1747–1819), Professor of Natural Philosophy at Edinburgh, attempted a wholesale rescue operation with his *Illustrations of the Huttonian Theory of the Earth* (1802). He sets out the problem of changing land/sea levels first in Section 3, § 36 (pp. 40–41)

'[W]e are . . . to inquire, from what cause . . . [the strata] from being covered by the ocean . . . are at present raised in many places fifteen thousand feet above the surface. Whether this great change of relative place can be best accounted for by the depression of the sea, or the elevation of the strata themselves, remains to be considered.

Of these two suppositions, the former, at first sight, seems undoubtedly the most probable, and we feel less reluctance to suppose, that a fluid, so unstable as the ocean, has undergone the great revolution here referred to, than that the solid foundation of the land have moved a single fathom from their place. This, however, is a mere illusion. Such depression of the level of the sea as is here supposed, could not happen without a change proportionally great in the solid part of the globe . . .'

Finally, 'We are to suppose, that the power of the same subterraneous heat, which consolidated and mineralised the strata at the bottom of the sea, has since raised them up to the height at which they are now placed, and has given them the various inclinations to the horizon which they are found actually to possess' (Section 47, p. 55).

Playfair returns to his argument in his Note XXI 'Changes in the apparent Level of the Sea' (1802, pp. 441–459) which gives a remarkably clear and modern view of the problems associated with what we would term eustatic or isostatic changes. 'To make the sea subside 30 feet all round the coast of Great Britain, it is necessary to displace a body of water 30 feet deep over the whole surface of the ocean. The quantity of matter to be moved in that way is incomparably greater than if the land itself were to be elevated; for though it is nearly three times less in specific gravity, it is as much greater in bulk, as the surface of the ocean is greater than that of this island' (p. 446).

But, as we have seen, all Wernerian views required the *sea* to have changed its level on one single occasion, and even the non-standard ideas of Buffon had a similar requirement. The only observable force known to the scientists of 1800 able to change the level of the *land* was that of an earthquake. Surprisingly, perhaps, Playfair does not discuss this explicitly, although the great Lisbon earthquake of 1755 had been a relatively recent and extraordinarily powerful event. I suspect one reason for the silence was (see Lyell, 1830, Vol. 1, pp. 438–40) that the predominant effect at Lisbon was of subsidence, not uplift. Even if one admitted that an earthquake might raise land by several feet (as well as sink it), it would require a huge number of such actions to produce the 15,000 ft elevations of land in the Huttonian system. So one came back to the basic sticking point: was there enough time? and were causes now in operation – such as earthquakes – sufficient to account for rocks and relief?

Playfair continues by showing that there is evidence that sea level has, during historical times, fallen in the Baltic (p. 445) and *risen* in the Mediterranean (pp. 447–50): 'it is evident that this cannot happen by the motion of the sea itself. The parts of the ocean all communicating with one another, cannot rise in one place and fall in another' (p. 451). What was to become a key piece of evidence of Lyell's armoury was the inferred recent history of the Temple of Serapis, at Pozzuoli (or Puzzuoli) (Figure 3.2) which Breislak had described as having its pavement underwater 'though it cannot be supposed that this edifice when built was exposed to the inconvenience of having its floor frequently under water' (Playfair, 1802, p. 450).

I hope I have made it clear that views of the earth science community of Europe and North America by the early nineteenth century, were in a pretty fair

Figure 3.2   *The Temple of Serapis, Pozzuoli. (From Lyell's* Principles, passim.)

muddle. There was still widespread uncertainty as to which parts of the landscape puzzle had to fit together, and which belonged to separate puzzles entirely. The great French palaeontologist, Cuvier, who (as we have seen) had been scathing about most seventeenth and eighteenth century *Théories* nevertheless produced his own, with successive cataclysms called to account for the changing suites of fossils (cf. Cuvier, 1825). The only major attempt to simply apply Huttonian principles to what was to become a classic site, the French Massif Central, was made by the English amateur George Julius Poulett Scrope (1797–1876). Scrope's work demonstrated clearly the repeated operation of volcanic outbursts and fluvial incision, exactly as Hutton and Playfair would have predicted. Yet his two great volumes (*Considerations on Volcanoes*, 1825; and *Memoir of the Geology of Central France* 1827 (although actually written in 1822: 1827, p. ix) received little attention. (See his later, 1858, extension of the earlier work.)

Whatever his defects, it was Charles Lyell's (1797–1875) Herculean achievement to cut through the welter of conflicting theories and observations and create a generally seamless and, above all, modern, scientific account of geological (and geomorphological) matters.

CHARLES LYELL'S 'PRINCIPLES' (FIRST EDITION, 1830–33)

Charles Lyell was a Scot, who spent most of his life in England. He trained as a barrister and practised Law, on and off, in a rather desultory fashion, until 1827. He was knighted in 1848, became a baronet in 1864 and was the first Professor of Geology at King's College, London (1831–3). Despite extremely poor eyesight, he was an almost compulsive field worker and, also, writer (see Chorley *et al.*, 1964, chapter 11; Wilson, 1972, 1999). He has always seemed to me an extraordinarily difficult figure to pin down. Perhaps because of his legal training, he is capable of shading his prose in ways which produce impressions at variance with ostensible meanings. He was, in many senses, overcautious: the North American term 'pussyfooter' comes to (my) mind, especially in his dealings with land ice, the work of rain and rivers (and evolution). Whether this was the manifestation of the legal mind; a wish to maintain decorous views which would fail to shock Victorian society; or merely an unduly timorous character, I am unable to judge. It is also the case that *The Principles* went through 11 editions during his lifetime and it has so far been more than any scholar can face to develop a true *variorum* edition.

That said, the first edition of *The Principles* (see the Penguin edition, edited by Secord, 1997) was a major and masterly exposition of a single-minded interpretation of the root matters of geology. It is worth giving a brief summary of the three volumes.

VOLUME I (1830; SECOND EDITION, 1832), VOLUME II (1832)
AND VOLUME III (1833)

Volume I opens with a quotation from Playfair (given above, Chapter 2, p. 20) concerning the uniform Economy of Nature and a frontispiece engraving of the Temple of Serapis (Figure 3.2). The first chapter ends with a brief recognition of Hutton's importance, thus:

> 'Hutton [was the first] who declared that geology was in no ways concerned with questions as to the origin of things. But his doctrine on this head was vehemently opposed at first, and although it has gradually gained ground, and will ultimately prevail, it is as yet far from being established' (1830, p. 4).

Rudwick (1970) gives an Analytic Table of Contents of the whole workwhich it is worth repeating here (Table 3.1). From this, it is evident that Lyell has separated out the 'modern causes' from the 'origin of rocks' and has also devoted virtually

Table 3.1 *Analytic table of contents of the first edition of Lyell's principles. (From Rudwick, 1970, pp. xi–xii.) The numbers in parentheses refer to chapters in the three volumes; headings in brackets are digressions from the main argument.*

SCIENTIFIC METHOD IN GEOLOGY (Vol. I, 1830)
   Introduction: the nature of geological science (1)
   Historical retrospect: the baneful influence of 'Mosaic' geology (2–4)
   The methodology of actualism (5)
THE FALLACIOUS BASIS OF PROGRESSIONISM
   Progressive climatic change contraverted
   The reality of climatic change (6)
   A uniformitarian theory for it (7)
   The theory tested and confirmed (8)
   Progressive organic change contraverted (9)
ANALYSIS OF PROCESSES NOW IN OPERATION
   Inorganic processes now in operation
   Aqueous processes
      Action of running water (10–14)
      Action of tides and currents (15–17)
   Igneous processes
      Volcanic activity (18–22)
      Earthquakes and movements of the earth's crust (23–26)
   Organic processes now in operation (Vol. II, 1832)
   The reality of the species-unit
      Lamarck's transmutation theory contraverted (1–2)
      The limits of variability (3–4)
   Geographical distribution of species, and means of dispersal (5–7)
   Theory of piecemeal ecological extinction of species (8–11)
   Effects of organic processes on inorganic
   [Organic processes not a counterpoise to erosion (11)]* misprint (12)
   Conditions for the preservation of organisms as fossils
      Preservation of terrestrial deposits (13–14)
      Preservation of terrestrial species in subaqueous deposits (15–16)
      Preservation of freshwater and marine species (17)
   [Coral reefs and the formation of limestones (18)]
RECONSTRUCTION OF PAST EARTH-HISTORY (Vol. III, 1833)
   Principles of reconstruction
   Actualism the only heuristic key to the past (1)
   The stratal succession and its discontinuity (2–3)
   Criteria of geochronology (4–5)
   Reconstruction of the Tertiary epoch
   Newer Pliocene period
      Sicily the link between present and past (6–9)
      Other areas (10–11)
   Older Pliocene period (12–14)
   Miocene period (15–16)
   Eocene period (17–20)
   [Tertiary strata not suddenly elevated (21–22)]
   Reconstruction of earlier epochs
   Secondary strata amenable to same methods (23)
   [Elevation of mountain chains not sudden (24)]
   The so-called Primary rocks: no vestige of a beginning (25–26)

the whole of Volume II to a consideration of the organic world (including – though apparently omitted by Rudwick – in Chapter XII, a discussion of what we would now term 'biogeomorphology' (see Viles, 1988).

I think there are three noteworthy elements about this first edition. First, Lyell – following pretty explicitly Cuvier (e.g. 1810, 1825) – is adamant that he is *not* providing yet another Theory (or *Théorie*) but rather a rational modern and scientific account of basic processes.

Second, he devotes a great part of Volume I to the demonstration that 'causes now in operation' are sufficient – given a Huttonian timescale – to create and perpetually modify global relief. Third, he explicitly develops what we would term a 'steady state' vision of earth surface processes, which is particularly clearly displayed in Volume I's discussion of climate change. It is useful to consider each of these in more detail.

NOT A THEORY, BUT BASIC, SCIENTIFIC PRINCIPLES

Although, as we have just seen, Lyell opens his Volume I with a quotation from Playfair and with a direct tribute to Hutton, he nevertheless rather buries explicit Huttonian views in his discussion of past approaches to matters geological. This occupies the Chapters II–IV and is a roll call of the major, but flawed theories from Oriental Cosmogony to the anti-Huttonian views of Kirwan and de Luc. This procession of what we might term 'hopeful monsters' (echoing Lamarck) is terminated by reference to three 'modern' and 'concrete' and 'progressive' developments. William Smith's geological map of England, completed in 1815, which remains 'a lasting monument of original talent and extraordinary perseverance' (p. 70: see Winchester, 2001); the foundation of the Geological Society of London, in 1807, whose proposed object was, not to theorize, but '[t]o multiply and record observations, and patiently to await the result at some future period' (p. 71); and finally, the methodological significance of the growing recognition of fossil sequences, due especially to the great French workers (he does not name them here, but Cuvier, Brogniart, Desmarest and Lamarck were of particular importance). 'The growing importance . . . of the natural history of organic remains, and its general application to geology, may be pointed out as the characteristic feature of the progress of the science during the present century' (p. 73).

This emphasis is probably most explicit in Volume III (1833), which takes direct issue with Werner's 'universal ocean' (pp. 37–38) and then lays out what is, in all essence, a modern picture of not just the origin of rocks, but also the methods by which their relative ages may be established. These – the 'law of superimposition' – and the comparison of fossil contents, remained the only way of giving relative geological chronology until the development of a whole battery of radiometric dating techniques from the early twentieth century onwards.

Although Lyell admits (1830, p. 71) that 'the reluctance to theorise was carried somewhat to excess' in the early nineteenth century, as a reaction to the bitter and sterile debates between upholders of Huttonian and Wernerian paradigms, he himself is remarkably cautious of quoting any 'evidence' which may smack in the slightest of a non-observable cause . . .

CAUSES NOW IN OPERATION

This – especially in Volume I – is where Lyell tries to overwhelm any scepticism on the reader's part concerning the ability of observable processes to produce observed topographic effects. In order to avoid, as far as possible, any remote suggestion that he is purveying secondhand 'traveller's tales'. Lyell goes to quite extraordinary lengths to demonstrate that *he* himself vouches for the truth of the phenomena described. (This is a most important characteristic of subsequent editions, where the results of his North American travels, in particular, take pride of place: Lyell, 1845, 1849; Wilson, 1999; Kennedy, 2001.) A striking instance of this comes in his discussion of the Temple of Serapis at Puzzuoli (Figure 3.2) and its significance: which is extensive (pp. 449–459) and accompanied by two further illustrations.

He begins *'Temple of Jupiter Serapis.* This celebrated monument of antiquity affords, in itself alone, unequivocal evidence, that the relative level of land and sea has changed twice at Puzzuoli, since the Christian era, and each movement both of elevation and subsidence has exceeded twenty feet.' This is followed by a personal account of 'a geological examination of the coast of the Bay of Baia . . . (as) we coast along the shore from Naples to Puzzuoli we find . . .' 'The sea encroaches as these new incoherent strata . . . when I visited the spot in 1828 . . . and I collected . . .' But that is not enough, to establish the *bona fides.* Lyell quotes no fewer than eleven authorities to support his own field observations. The Temple, briefly, shows evidence for up and down motions, which can be firmly linked to dated earthquakes: as Playfair noted, Breislak had pointed out that the Temple originally must have stood above sea level. Its floor is currently awash at 1 ft below high water mark (and a lower floor has been found below the present), suggesting the sea level has risen at least twice. But, most strikingly, the marble – but not the granite – columns of the temple are found to contain the boreholes made by a range of extant marine organisms, to a height of some 23 ft above high water mark. So the Temple must have sunk by at least 22 ft and then risen again, all within the historic period.

Lyell uses the Temple of Serapis not merely as a frontispiece in his first volume, but, embossed in gold, on the cover of later editions. His 1830 discussion encapsulates his method of demonstrating 'the sufficiency of modern causes': if one part of the Mediterranean coastline can be *proved* to have gone up and down by more than 20 ft in 2000 years, then raising beds of shells 15,000 or 20,000 ft in the Andes is clearly a feasible proposition, always given sufficient time.

What is also of interest in Lyell's discussion of tectonic and subaerial forces, is the weight he gives to the different agencies (see Table 3.2).

This should dispel any idea that all Lyell's causes now in operation were thought to produce 'gradual' changes. In fact, if we examine the account of the erosional work of running water, we find a substantial proportion (pp. 191–197) is devoted

Table 3.2 *Attention paid to 'present causes' (Lyell, 1830, Vol. I)*

AGENT	CHAPTER(S)	PAGES
Running water (including deltas)	X–XIV	168–255
Tides and currents	XV–XVII	256–302
Volcanoes	XVIII–XXII	314–397
Earthquakes	XXIII–XXVI	398–460

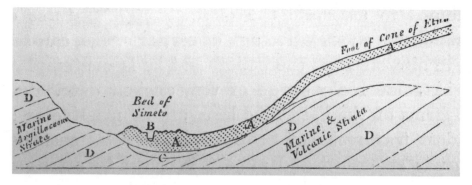

Figure 3.3 *'The power of running water . . . ?' Lyell's diagram of the River Simeto.*

first, to the impact of the New Madrid earthquakes of 1812 on the course of the Mississippi and, second, to a catalogue of cataclysms relating to 'Floods, bursting of lakes, etc.' The direct evidence of what we might term 'normal' fluvial erosion is on the thin side (Figure 3.3): the River Simeto has eroded the dated lava flow in 'about two centuries' and its passage is 'from fifty to several hundred feet wide, and in some parts from forty to fifty feet deep'. (It is worth noting that Scrope (1827) could provide far more striking cases from the Massif Central; Lyell (1830, p. 171) quotes a later paper (Scrope, 1830, published 1834) on valleys in general, but omits any evidence from the Auvergne.) However, more attention is paid to the 'reproductive' effects of running water, i.e. the amount of deposition associated with rivers and, in particular, the extent of deltas. These, of course, are evidence for subaerial erosion but also change the location of coastlines. By and large, a case could be made that Lyell's discussions of modern causes are very much focused on those which create changes at the coast: especially in elevation. His key agencies are earthquakes and marine action.

In many ways, Playfair's account of subaerial processes (summarized in the quotation at the head of this chapter) is a more appealing account to those of us who would agree about the general paramountcy of rain and rivers in fashioning global landscapes over 4000 million years. However, what neither Playfair nor Lyell could take into account, was the rôle of land ice (see Chapter 4) and the evidence of wholesale global climate change. This was to conflict with Lyell's third major strand of thinking.

THE STEADY-STATE EARTH

We have seen (above, p. 31) that Hutton was unclear about the exact replacement of old continent by new, although he felt sure that Divine Providence would always see to it that there was 'sufficient' land available for organisms.

Lyell took the notion of steady state far further. He considered that there was no evidence for *progressive* changes, either in the organic or the inorganic world. This was, indubitably, a viewpoint which derived from his desire to be completely distanced from all those theoretical constructs which saw a divinely ordained sequence of terrestrial (and organic) environments. However, there was ample evidence – which Lyell discusses at length, especially in Volumes II and III – both for very different depositional rock types and collections of organisms; reflecting substantial differences in climatic conditions. (Buffon, you will recall, defined his Sixth Epoch as that when tropical animals lived in the north of the Northern Hemisphere.) What machinery could Lyell adduce to account for these shifts in climate, within his steady-state vision?

The answer is set out, with great clarity, in Chapters VI–VIII of Volume I.

Buffon's warmer epoch was seen by him as proof of the continuous cooling of the globe: Lyell will have none of this.

'But if, instead of vague conjectures as to what might have been the state of the planet at the era of its creation, we fix our thoughts steadily on the connection at present between climate and the distribution of land and sea; and if we then consider what influence former fluctuations in the physical geography of the earth must have had on superficial temperature, we may perhaps approximate to a true theory. If doubts still remain, it should be ascribed to our ignorance of the laws of Nature, not to revolutions in her economy; – it should stimulate us to further research, not tempt us to indulge our fancies in framing imaginary systems for the government of infant worlds' (1830, p. 105).

The key is 'fluctuations in physical geography'. Changes in the organization of land and sea and in the location of mountains are the principal machinery to be employed. Although Lyell also notes (p. 104) that cosmogonists have invoked shifts in the Earth's rotational axis to explain climate change, and he accepts (p. 110) that this phenomenon termed the 'procession' (we would say 'precession') of the Equinoxes may change solar radiation receipts on a 10,000 year cycle.

But if, as he assumes (pp. 112–13), there is: always the same proportion of land and sea over the globe; always the same maximum and mean elevation of the land; always the same mean and maximum depth to the oceans; and always the same continental configuration . . . then what will control the climate (and therefore the subaerial processes and the life forms) at any location will be, first, its own elevation and distance from the sea and, second, the disposition of land and sea with respect to the poles and Equator. For Lyell does admit (p. 113) that continental configurations may change.

So: 'If this be admitted, it will follow as a corollary, that unless the superficial inequalities of the earth be fixed and permanent, there must be never-ending fluctuation in the mean temperature of every zone, and that the climate of one era can no more be a type of every other, than is one of our four seasons of all the rest' (1830, p. 115).

Lyell acknowledges the importance of the global links between climatic zones, such that '. . . no great disturbance can be brought about in the climate of a particular region, without immediately affecting all other latitudes, however remote' (p. 117).

But – and this has great significance in Lyell's views of Louis Agassiz's Ice Age – 'we may observe, in conclusion, that however great, in the lapse of ages, may be the vicissitudes of temperature in every zone, it accords with our theory that the general climate should not experience any sensible change in the course of a few thousand years, because that period is insufficient to affect the leading features of the physical geography of the globe' (p. 123). A statement of profound importance, as we shall see with respect to Lyell's reaction to the Ice Age (Chapter 4).

This section of the *Principles* illustrates both Lyell's huge strengths and his weaknesses. The mechanisms he puts forward for climate change – especially the importance of the precise configuration of land and sea and the location and elevation of major mountain ranges – are now held to be of great significance in climate shifts (as are alterations in the Earth's planetary motion; see Chapter 4). However, despite all his protestations, you will note that Lyell, too, has a theory (though not, of course, a Theory) and his unwavering attachment to a steady-state and non-progressive world was to create very serious difficulties. It did not accord with the great glacial vision of Agassiz (Chapter 4); and it sat most uncomfortably with the evolutionary ideas of Charles Darwin (see Chapter 5). But it was a copy of Lyell's Volume I which the neophyte Darwin was handed by Robert Fitzroy at the outset of the *Beagle* voyage and, in some ways, that was to prove one of Lyell's greatest triumphs.

The *Principles* was not only a magisterial tour de force, it was a modern outline which went far beyond Playfair and, since it had rejected both Wernerian views and other outdated Theoretical structures, it demonstrated marked superiority over the compendious visions of Jameson, De Luc, Breislak and even Cuvier. It was to run through eleven more editions (last 1875), with sometimes astounding absence of substantive changes (Chapters 5 and 6). It is impossible not to pay tribute to Lyell's devotion to the verifiable and to his widespread dissemination of up-to-date principles. But his native caution and his excessive reliance on his steady-state theoretical underpinning were to prove real drawbacks. As was his insular devotion to the power of the sea . . .

For the next great invention, that of The Ice Age, clashed not merely with Lyell's methodology, but proposed a rôle for land ice that far eclipsed marine forces.

Chapter 4

Inventing the Ice Age: the Rôle of Louis Agassiz

'Every river appears to consist of a main trunk, fed from a variety of branches, each running in a valley proportioned to its size, and all of them together forming a system of valleys communicating with one another, and having such a nice adjustment of their declivities, that none of them joins the principal valley, either on too high or too low a level; a circumstance which would be infinitely improbable if each of these valleys were not the work of the stream that flows in it' (John Playfair, 1802, p. 103).

CONTINUING PROBLEMS WITH HUTTON AND PLAYFAIR'S TOPOGRAPHIC VISION

The quotation given above is generally termed 'Playfair's Law'. However, despite a widespread belief that it accurately depicts the generality of fluvial valley networks, it patently does *not* fit the situation in much of northern and western Europe and northern and eastern North America. Just as Saussure demonstrated that Bourguet's valley *'angles saillans et rentrans'* were not in evidence in the Alps, so there were more than the – still inexplicable – piles of diluvium which provided ample evidence of the error of Huttonian ways. Amongst the key instances where the Huttonian theory certainly did not fit the facts were:

1. the non-accordance of valley junctions, especially in the Alps and other mountain ranges (Figure 4.1);
2. the widespread occurrence of deep lakes in the upper courses of valleys – Lac Leman (the Lake of Geneva) was an especially unyielding example of a rock basin which could not have been excavated by the 'normal' action of rivers;
3. the existence of cases – such as those of the Cotswolds – where the stream was minuscule compared with the size of the valley;
4. worse, the widespread occurrence of valleys with no streams in them at all;
5. the existence of valleys – such as the fjords of Scandinavia – which patently continued out under the sea, yet rivers could not erode below sea level;
6. those widespread deposits of non-local sands, gravels and boulders – Buckland's Diluvium;
7. the fact that rivers not infrequently ran in valleys which cut dramatically through high ground – key examples existed on the flanks of the Wealden dome (see Chapter 5), but an even more dramatic example has been described by Thomas Jefferson (1788, p. 17).

(A)

(B)

Figure 4.1 *Hanging valleys, New Zealand.*

(C)

Figure 4.1   (*cont'd*)

'The passage of the Patowmac through the Blue Ridge is perhaps one of the most stupendous scenes in nature. You stand on a very high point of land. On your right comes up the Shenandoah, having ranged along the foot of the mountains an hundred miles to seek a vent. On your left approaches the Patowmac, in quest of a passage also. In the moment of their junction they rush together against the mountain, rend it asunder, and pass to the sea.'

Once again, this list, we would now think, contains questions whose answers do not all come from one source. What we shall be concerned with in this chapter are the first five objections: we shall leave a discussion of the sixth until the following chapters.

It must be admitted that all the first five objections were extremely well-founded and serious. Although they gave huge comfort and glee to those fervent adherers to Moses and/or Werner such as Jameson and de Luc, they were quite unanswerable within the Huttonian framework. One of the most painful passages in Playfair's *Illustrations* comes when he tries to find a Uniformitarian origin for the Lake of Geneva (Note XVI, especially #321–326). His best effort is to invoke the existence of some soluble strata covered by a clay layer and 'Should this covering be broke open by any natural convulsion, or should it be worn away, as it must be by the general progress of the detritus, the water would gain admission to the saline strata, would gradually dissolve them, and form of course a very deep and extensive lake, where all was before dry land. This event is not only possible, but it should seem, that in the course of things it must necessarily happen.

'326. Something of this kind may have taken place in the track of the Rhone, and may have produced the Leman Lake' (1802, pp. 364–5).

One can only wince.

It should be evident that the only answers to the principal objections were going to come from the invocation of some new agency. We now think that land ice was the solution.

OBSERVATIONS ON GLACIERS

'For the moving of large masses of rock, the most powerful engines without doubt which nature employs are the glaciers, those lakes or rivers of ice . . .' (Playfair, 1802, p. 388).

The 'sublime' scenery of high mountains had become fashionable in the late eighteenth century and soon no educated person's Grand Tour would be complete without a personal encounter with at least one Alpine glacier (generally the Mer de Glace at Chamonix). Saussure's huge, four-volume catalogue of his Alpine travels (1779–1796) left no-one in any doubt as to the nature of mountain scenery and the setting of glaciers amidst huge volumes of often non-local rocks and boulders. It was well known that the snouts of valley glaciers fluctuated in their location and it was perfectly evident that they could transport huge blocks. ('These fragments [of rock] they gradually transport to their utmost boundaries, where a formidable wall ascertains the magnitude, and attests the force, of the great engine by which it was erected': Playfair, 1802, p. 389.)

It seems clear (see the discussion in Chorley *et al.*, 1964, chapter 13) that it was the recognition of similar 'formidable walls' – we would say 'terminal moraines' – at some distance down valley away from the present glacier snout that began to give Swiss residents the idea of former extensions of glaciers. First a guide, Perraudin, then Venetz (a professional engineer) and finally Jean de Charpentier (1786–1855) the Director of Mines of the Canton of Vaud not merely envisaged extending the length, but also the depth and breadth of the existing valley glaciers. This, of course, helped to explain the evidence of diluvial dumping, although not necessarily the presence of non-local rock types (which we call 'erratic blocks'). But evidence was also being presented for the active grinding and polishing of surfaces in these once-glaciated valleys and, in particular, the striation of rocks.

Although of interest, the local advance and retreat of valley glaciers was not a wholly revolutionary concept: every Grand Tourist knew about it. As Cunningham states: 'James David Forbes' visits to the Mer de Glace in 1827, 1832 and 1839 had been merely day trips familiar to tourists. How are to be explained the facts that in spite of extensive Alpine journeys up to 1841 Forbes evinced no scientific interest whatsoever in glaciers, yet in 1842 emerged as the leading British glaciologist? In a word the answer is Agassiz' (Cunningham, 1990, p. 109).

But before we look at Agassiz and his invention, we need to outline modern thinking about the causes of climate change (recall Lyell's views, Chapter 3).

WHY CLIMATE CHANGES

There appears ample geological evidence for major and repeated wholesale changes in the Earth's climate – those widespread tropical swamps now represented by the Carboniferous coal measures, or the evidence for a vast extent of warm shallow seas in which the Chalk of the Cretaceous was laid down, or, in contrast, the wholesale development of desert sandstones during the Permo-Triassic – all seem to bear out Lyell's warning that 'the geologist should not hastily assume that the temperature of the earth in the present era is a type of that which most usually obtains' (Lyell, 1830, p. 106). The present mean temperature of the Earth's surface is some 13 °C (Lovelock, 1988, p. 9): that is, we feel certain, much cooler than the situation for most of geological time. Why are there such variations?

We consider that the Earth's surface is heated by the receipt of short-wave solar radiation and that it is the subsequent emission of long-wave radiation from the surface which heats the atmosphere. Wholesale changes in global temperature may, therefore, be the product of both extraterrestrial and terrestrial causes. The most fundamental cause is variation in the output of solar radiation, both absolutely and in terms of wavelength: this may – on the short term – be linked to the long-observed sequence of high and low sunspot frequency. Second, the Earth as an astronomical body shows three sources of cyclic variation in its relation to the sun's rays: in the tilt of its axis; in the form of its elliptical path about the sun ('the obliquity of the ecliptic'); and – as Lyell realized – in the precession of the equinoxes, when the portion of the Earth's surface most directly exposed to the sun's rays varies. Then there is – as the general public of the late twentieth and early twenty-first centuries is all too agitatedly aware – variability in the gaseous composition of the atmosphere as a result of volcanic activity, fire and vegetation changes, as well as human malfeasance, which alters both how much radiation is received at the surface and how much outgoing heat is trapped. The surface itself may vary both in the proportion of the incoming radiation it reflects (its albedo), which is trapped by the natural or the enhanced 'greenhouse effect', and in its capacity to absorb that radiation and emit heat. Snow cover alters albedo seasonally.

Land and sea have very different albedos and thermal characteristics. So in an important sense Lyell was correct, since we now think that the changing configuration and global positioning of the continental masses will have had enormous impact on the world's climate. Finally, there is the rôle of high mountains: these are the site of very intense mechanical and chemical weathering and, we now consider, this dramatically affects the carbon dioxide content of the atmosphere and hence its temperature.

All of this not merely adds up to great complexity, but it also involves causal mechanisms which may have different degrees of predictability and which certainly work on different timescales. In our current thinking the onset of periods which see the repeated wholesale development, advance, retreat and disappearance of great ice sheets – and there seems to be evidence for at least four of these in the geological record (see Bowen, 2004) – is probably linked to events produced by plate tectonics and therefore occurs more or less at random. But once an Ice Age is established, the alternation of icy and temperate episodes (and we think we can see evidence for as many as 50 in the last 2.5 million years, Bowen, 2004, p. 550) appears to be strongly linked to the variability in the three, linked astronomical cycles (tilt, precession and obliquity) whose significance was first put forward by the Scot, James Croll (1821–1890), but whose mechanics were elaborated by the Serbian mathematician, Milutin Milankovitch) (see Imbrie and Imbrie, 1979; Bryson, 2003, pp. 375–8). Within each glacial or interglacial, there may be short temperature fluctuations – sometimes extremely abrupt – that are increasingly being ascribed to the behaviour of the great ocean currents (cf. Broecker, 2000, 2003, 2004), but – on a smaller scale – can also be due to huge volcanic events, such as the eruption of Krakatau in 1883 (see Thornton, 1996).

It is by no means clear how the climate system works, especially over long time periods. But it is evident (cf. Barry, 1997) how intimately the state of climatic variables affects geomorphological and geological processes.

JEAN LOUIS RODOLPHE AGASSIZ (1807–1873)

One of the greatest revolutions in our view of the Earth was brought about by a young, pious Swiss expert on the biology of fishes (Lurie, 1960). Louis Agassiz studied in Zürich, Heidelberg, Munich and Erlangen, and then went to work with Cuvier in Paris, before becoming Professor at Neuchâtel in 1832. It was from here, that he made the expedition which established his great theoretical vision of *Die Eiszeit* (The Ice Age): see later. In 1846 he moved to the USA and, from 1847 until his death, held a chair in Zoology at Harvard. His later years were marked by a vision of the world as being driven by great glacially determined revolutions, which harked back to Cuvier's ideas. There was no room for Darwinian evolution. And there was an extension of glacial explanations even to the mouth of the Amazon, where he confidently identified a large terminal moraine during the Thayer Expedition to Brazil (1865–6): see Hartt (1870). He was certainly a man with a Ruling Hypothesis.

But his chief contribution to our view of our planet was his vision of a Northern Hemisphere covered by a great sheet of ice. This vision was formulated after Agassiz paid a visit to Chamonix and Diablerets with de Charpentier in 1836, and was announced by Agassiz to the Helvetic Society in 1837 (see Carozzi, 1967), where the reception was distinctly cool. The great Humboldt wrote expressing his unease at the whole idea and, by and large, that was the general reaction of the contemporary Continental earth scientists (see Chorley *et al.*, 1964, pp. 205–6). But Agassiz persisted and, in 1840, his great – and superbly illustrated (see Figure 4.2A) – *Etudes sur les glaciers* was published (see Carozzi, 1967).

In this, Agassiz drew on a whole range of topographic evidence to convince. First, there was no doubt that valley glaciers moved. Second, there was no doubt that they could carry enormous loads of sand, gravel and rocks of all sizes. Third, as Saussure had noted, the lines of lateral moraines and the trains of gravel extended for long distances from the Alps to the Jura and from the Alps to the Mediterranean: points Playfair had picked up, but for which he had been unable to provide a satisfactory explanation (1802, pp. 381–388). Fourth, as Figures 4.2A and 4.2B show, the area around an extant glacier was marked, not just by erratics of all sizes, but – less spectacularly but more significantly – by rock surfaces scratched and polished in ways that seemed uniquely related to the agency of ice. Certainly it seems that this last phenomenon was crucial in converting the Diluvialist Buckland to Agassiz's camp: Buckland visited Agassiz in Switzerland in October 1838 and his attention 'was . . . directed . . . to the phenomena of polished, striated and furrowed surfaces on the south-east slope of the Jura, near Neuchâtel' (Buckland, 1840–1, p. 332). (However, Buckland did not entirely retain his convictions: see Chorley *et al.*, 1964, p. 222.)

Despite the resistance to his ideas on the Continent – and despite, too, the fact that there were some very serious flaws in their development, such as the notion that the great sheet of ice antedated the uplift of the Alps; and the minor point that valley glaciers actually flow faster in the middle than at the edge, rather than the reverse (a point he quite rapidly corrected) – Agassiz used a visit to the British Academy meeting in Glasgow in September 1840 to promulgate his vision. He went on, after a tour of Scotland and Ireland, to a November meeting of the Geological Society in London. In the process, he converted Buckland, convinced Darwin, failed to impress Sir Roderick Impey Murchison (1792–1871) and – for one brief moment – even persuaded Lyell that it was land ice, not icebergs drifting in on raised sea levels, that was responsible for key elements of Scotland's scenery. The core of his vision, as presented in London, was as follows:

'If the analogy of the facts which he has observed in Scotland, Ireland, and the north of England, with those in Switzerland, be correct, then it must be admitted . . . that not only glaciers once existed in the British Islands, but that large sheets (nappes) of ice covered all the surface.

It must then be admitted . . . that great sheets of ice, resembling those now existing in Greenland, once covered all the countries in which unstratified gravel is found; that this gravel was in general produced by the trituration of the sheets of ice upon the subjacent surface, that moraines . . . are the effects of the

(A)

(B)

Figure 4.2 *(A) A glacial valley with moraine, erratics and polishing and striations. (Agassiz, from Carozzi). (B) Lyell's view of a glacial snout.*

retreat of glaciers; that the angular blocks found on the surface of the rounded materials were left in their present position at the melting of the ice; and that the disappearance of great bodies of ice produced enormous debacles and considerable currents, by which masses of ice were set afloat, and conveyed in diverging directions, the blocks with which they were charged' (Agassiz, 1840–1, p. 331 quoted in Chorley et al., 1964, p. 215).

As the biologist Edward Forbes (1815–1854) was to say in a letter to Agassiz, in February 1841: 'You have made all the geologists glacier mad here, and they are turning Great Britain into an ice-house.' (quoted by Cunningham, 1990, p. 59).

THE SIGNIFICANCE OF AGASSIZ'S IDEAS

First and foremost, the new vision disposed of the increasingly embarrassing diluvium. Second, as we noted in Chapter 1, Ramsay (1862) would show it disposed of the problem of lake-filled rock basins (as well as the absence of Bourguet's *'angles saillans et rentrans'* in mountain areas). Third, it showed that fjords could be seen as carved by glacier extensions. Fourth, it provided an obvious origin for the non-accordant valley junctions of mountainous regions: the 'hanging valleys' (Figure 4.1) representing the former sites of minor glaciers.

But it was in the broader ramifications of The Ice Age vision that greatest importance resided.

First, if there was all that water locked up in huge ice sheets, then sea level must have been depressed. The concept of eustasy was, in fact, put forward as early as 1842, by the Scot Charles McLaren (1782–1866), although a full working-out was delayed until 1865 (Davies, 1968, p. 300) and the work of Searles Valentine Wood, Junior (1830–1884). Second, the size of the ice sheets represented a substantial load upon the land and in 1865, Thomas Francis Jamieson (1829–1913) suggested the concept of isostasy be applied to the loading and unloading of the land surface by the great ice sheets as they waxed and waned (Davies, 1968, p. 299). These two mechanisms – eustasy and isostasy – combined go a long way to explain some of the more puzzling manifestations of changing levels of land and sea. The joint consequences of these forces are, however, by no means straightforward to unravel and, as the number of glacial advances and retreats which have been identified in the past 2–3 million years has been multiplied, the unravelling of the concomitant eustatic and isostatic effects (plus, of course, more straightforward tectonic inputs) has become something of a nightmare.

Second, if there had been wholesale refrigeration of the Northern Hemisphere, then – since climatic conditions in one zone, as Lyell had noted in 1830, influenced those in others – it must have been cooler and possible wetter (or drier) in the regions beyond the ice sheets. The concept of a periglacial – or ice margin – zone was to emerge in which the dry valleys and 'underfit streams' (see Dury, 1964–5) so worrying in the simple Huttonian view could be related to the existence of seasonal or perennially frozen ground.

Third, Agassiz's Ice Age had come and gone: it is worth noting that his mentor, Charpentier, had reasoned that it postdated the Alpine uplift, but his book – *Essai sur les glaciers* (1841) – was eclipsed by Agassiz's ebullience (Chorley *et al.*, 1964, p. 210). This could in no way be easily reconciled with Lyell's steady-state vision, in which parts of the Earth's surface underwent warming or cooling, yet the whole planet somehow retained its present mean character. The invention of the concept of global, *reversible* climate change, associated firmly with Agassiz, has become one of the key notions of our modern earth science even though we are far from certain about the machinery involved.

The implications of Agassiz's vision were, truly, staggering. Even after more than 160 years of research, we still are very far from a complete picture of what we term the Pleistocene glacial period. So profound were these implications that it is scarcely surprising that they met with resistance. In the absence of our whole suite of ideas about the multiple causes of climatic oscillations (as outlined above) and given the very imperfect methods for dating both deposits and erosional features such as cirques and raised strand lines (such as the Parallel Roads of Glen Roy – see below and Chapter 5), Agassiz's vision smacked remarkably of some supernatural debacle. Was it really within the scope of the workings of 'causes now in operation', or did it just substitute an icy for a watery source of the disturbed rocks and relief of much of the Northern Hemisphere?

LYELL'S REACTIONS

Lyell, frankly, first back-tracked and then wavered (see Chorley *et al.*, 1964, pp. 232–3). It is interesting to see the oscillations in his views, as set out in the *Principles*.

In the 1830 first edition, Volume I, chapter VI opens with evidence for the former *warmth* of the Northern Hemisphere, including (pp. 96–97) an attempt to explain the permafrost-bound and iceberg-encased corpses of mammoths in Siberia. But the explanation was (as shown above, Chapter 3) non-revolutionary. 'Siberia and other arctic regions, after having possessed for ages a more uniform temperature may, *after certain changes in the form of the arctic land* [my emphasis] have become occasionally exposed to extremely severe winters . . .' (pp. 97–98). This is followed by a straightforward account of the way glaciers form and explicit recognition that they move: 'the body of a glacier which slowly descends to the sea, and becomes a floating iceberg' (1830, p. 98). A brief account of the local transporting power of Alpine glaciers and the form of what we term lateral moraines is then given (pp. 175–176).

Lyell produced a three-volume 6th edition in June, 1840 – too early for the Agassiz revelations at Glasgow yet Lyell had caught the icy wind of change and begins a clear discussion of glacial movement (Vol. I, pp. 375–7) with the statement 'It has been a generally received opinion' says M. Agassiz 'from the time of Saussure . . .' There is also (p. 377) an acknowledgement of Charpentier's description of moraines 'as entirely devoid of stratification, for there has been no sorting of the materials, as . . . when deposited by running water. The ice transports indifferently, and to the same spots, the heaviest blocks and the finest particles, mingling all together, and leaving them in one confused and promiscuous heap wherever it melts.'

Clearly, Lyell is quite excited by these new ideas, but there is a characteristically nuanced termination to the discussion: 'To conclude; it appears that large stones, mud and gravel, are carried down by the ice of rivers, estuaries, and glaciers, into the sea, where the tides and currents of the ocean, aided by the wind, cause them to drift for hundreds of miles from the place of their origin' (1840, Vol. 1, p. 381). This statement was succeeded by a lengthy illustrated discussion of the rôle of the seasonal ice which fringes northern coasts in the transport of erratics.

By the appearance of the one volume Seventh Edition in 1847, there is little added, with the exception of a moderately detailed account of the way in which glacial striations are produced, based on Agassiz's November 1840 paper (Agassiz, 1840b; see Davies, 1968, pp. 281–4).

But, by the one volume 9th edition of June 1853, there is some shift. An acknowledgment of evidence for 'a colder, or as it has been termed "glacial epoch", towards the close of tertiary periods throughout the northern hemisphere' (1853, p. 75), although Lyell feels that this can be explained by topographic changes, especially by the rise of land in northern Russia, which he thinks made the area more continental and, hence, colder. (Some of that rise we would attribute to the reverse cause, namely the isostatic recovery of the areas after the ice sheets melted and the climate warmed.) There is no land ice in his discussion of the 'glacial epoch' and its possible causes. There is no real acknowledgement of land ice in Chapter XV, entitled 'Transportation of solid matter by ice' (1853, p. 219). Lyell's conclusion is: 'The agency of glaciers in providing permanent geological changes consists partly in their power of transporting gravel, sand, and huge stones to great distances, and partly in the smoothing, polishing and scoring of their rocky channels, and the boundary walls of the valleys through which they pass' (1853, p. 226).

But as for icebergs and lake ice (Figure 4.3), now . . . 'There can be little doubt that icebergs must often break off the peaks and projecting points of submarine mountains, and must grate upon and polish their surface, furrowing and scratching them in precisely the same way as we have seen that glaciers act on the solid rock over which they are propelled' . . . (1853, p. 229). And, if you want further material – on icebergs, of course – the footnote refers you to the accounts of his two North American journeys (1845, 1849) and his *Manual of Geology* (4th edn, 1852) . . .

By the 10th edition (1867) a major change is introduced by Lyell's need to deal with Croll's theories about the astronomical causes of reduced radiation, to which a whole chapter (XIII. Vicissitudes in climate how far influenced by astronomical changes, pp. 268–304) is devoted. To say Croll's ideas are fought tooth and nail is putting it mildly. It is particularly instructive to watch how Lyell deals with Croll's attempts to identify substantial minima in solar input (p. 293). The two most recent of these Croll considered would have occurred about 100,000 and 200,000

(A)

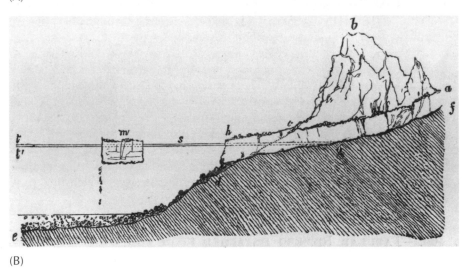

(B)

Figure 4.3   *How to move rocks by ice. (A) Lake ice (Lyell). (B) Icebergs (Geikie).*

years before 1800. 'The periods . . . would not, I conceive, be sufficiently distant from our era to afford time for that series of glacial and post-glacial events which we can prove to have happened since the epoch of the greatest cold' (Lyell, 1867, p. 294) . . . Why not? Well, Lyell is clinging to topographic changes – now seen as alternating between the northern and the southern hemispheres – as the cause of the periodic refrigeration and warming and – as noted above, p. 40 – he simply cannot provide enough vertical displacement within Croll's timeframe . . .

At the root of Lyell's problems are, we would think, that he got his ultimate and proximate causes of glacial epochs well and truly muddled. Our current view (see above) is that it is probable that it is the broad arrangement of land and sea and the location of major mountains which trigger periodic epochs of global refrigeration. Within each of these epochs, the role of astronomical forcing seems to govern the oscillations from cold to warm. Further, we are reasonably clear that cold, glacial episodes are marked by low sea level. Lyell wanted none of the astronomic causes: he basically does not accept that the possible variations which they could produce are adequate to freeze the Northern Hemisphere (or the Southern): in this, his caution may have been well-founded, given modern unease about the precise mechanisms which may be at work. However his topographic explanations are frankly nonsense: for one thing, he wanted *high* sea levels during the coldest periods.

'[G]reat oscillations in the level of the land since the commencement of the Glacial Period [are] proved to have taken place. The change of level in Scotland . . . amounts to 500 feet, in some parts of central England to 1200 and in North Wales to 1,400 feet . . .' (1867, p. 195). No wonder he was more than unhappy at trying to compress this scale of sea-level change into Croll's 100,000 or 200,000 years . . . If you could stretch things back to the Eocene, now . . . and Lyell provides a helpful map showing how much of Europe has patently been submerged in that (at that point, of course, unknown) longer interval . . .

In one further area, Lyell's ideas were actually valuable, but associated with misconceived temporal frameworks. Even in the 1853 edition, he was keen to link the variability in iceberg volumes in the North Atlantic with the climate of its shores: 'It was towards the close of "[the Pliocene]" that the seas of the northern hemisphere became more and more filled with floating icebergs . . . so that the waters and the atmosphere were chilled by the melting ice . . .' (1853, p. 86), and, again 'It is a well-known fact that every four or five years a large number of icebergs, floating from Greenland . . . are stranded on the west coast of Iceland. The inhabitants are then aware that their crops of hay will fail, in consequences of fogs . . . and the dearth of food is not confined to the land, for the temperature of the water is so changed that the fish entirely desert the coast' (1853, p. 97). The role of 'Iceberg Armadas' in giving extreme and sudden depression of northern mid-latitude temperatures, by 'drowning-out' the Gulf Stream, has emerged in the late twentieth century as a key mechanism in climate change: but the icebergs in our contemporaries' story are those emanating from the collapse of the great Laurentide ice sheet of North America, or the excessive melting of Arctic ocean ice . . . (see Roberts, 1998; Broecker 2000, 2003, 2004; Clarke *et al.*, 2003). However, the 10th edition of the *Principles* did see some crumbling of Lyell's scepticism about land ice. First, he had actually visited Switzerland twice (in 1857 and then in 1865) and, on the first occasion, was moved to accept the erosional evidence of the woodcut (Figure 4.2B) which he had already used:

'In the foreground . . . some dome-shaped masses of smoothed rock are represented, called in Switzerland "roches moutonnées" for they are compared to the backs of sheep which are lying down. These owe their rounded and smooth outline to the action of the glacier when it was more in advance, and the irregularities of the hard rock having been planed off' (1867, pp. 375–6). He had seen it for himself, so it was real . . .

And, finally, in 1865 he had visited the ice-dammed lake of the Märjelen See, which drained periodically with the dramatic floods we term *jökulhaups*. '[I examined] a point of great geological interest, namely, the form and structure of a large terrace or line of beach which encircles the lake basin all round its margin, and which constitutes its shore when it is full . . . I satisfied myself that this terrace is a counterpart of one of those ancient shelves, or parallel roads, as they are called, of Glen Roy in Scotland, which, as Agassiz first suggested, were probably formed on the edge of lakes dammed up by ice, which may have existed in the Glacial Period in Scotland' (1867, p. 377). So, 27 years on, Agassiz was publicly acknowledged as correct, though scarcely in the most enthusiastic terms and only in part.

All in all, Lyell epitomises the distinctly lukewarm acceptance of Agassiz's great vision, which Chorley *et al.* (1964) discuss in some detail. But I think Lyell's pussyfooting had some very particular roots. There was his congenital reluctance to accept anything other than the evidence of his own eyes. There was, in consequence, his fixation on the role of the sea and, in particular, of icebergs – rather than land ice – as the perpetrator of both erosion and the distribution of erratics (hence, by the way, our term glacial drift – coined by Murchison). And, underlying it all, there was his deep and profound belief in a steady-state Earth, where a limited number of 'causes now in operation' could be shown to have sufficient power to alter climate: although by 1867, he had accepted that the 'Glacial Epoch' consisted of two 'continental periods' (p. 195) or interglacials, he simply could not accept that any astronomical cause was sufficient to allow the development (and retreat) of huge sheets of ice. (It may be worth repeating that there is still some unease at the energetics of the Croll–Milankovitch mechanism, although the dating of glacial–interglacial oscillations seems to confirm its significance. When – or if – the banded Martian ice caps can be sampled and dated, it may be possible to test the theory with reference to the astronomical cycles experienced by that planet.)

The root of all Lyell's wavering was a view of the Earth that was both over-orderly and profoundly insular, in the sense that the ocean was seen as far and away the most potent of the aqueous agencies of denudation. One important convert to that view, with very interesting consequences, was the young Charles Darwin . . .

Chapter 5

Inventing a Balanced View of 'Forces Now in Operation' at the Earth's Surface: Charles Darwin's Travels in Space and Time

'I am tempted to give one other case, the well-known one of the denudation of the Weald . . . If . . . we knew the rate at which the sea commonly wears away a line of cliff of any given height, we could measure the time requisite to have denuded the Weald . . . I conclude that for a cliff 500 feet in height, a denudation of one inch per century for the whole length would be an ample allowance. At this rate, on the above data, the denudation of the Weald must have required 306,662,400 years; or say three hundred million years' (Charles Darwin, *On the Origin of Species*, 1859, pp. 285–87).

INTRODUCTION

The passage quoted above is generally seen as interesting because its Huttonian/Lyellian use of observations of present-day processes to infer the necessary lapse of geological time roused the ire of William Thomson, Lord Kelvin, with the consequences we have outlined in Chapter 2. However, it is also pertinent in the present context that Darwin assumes, without discussion, that the agency responsible for the Wealden Denudation was the sea. The present chapter will use Darwin's views of the processes now in operation at the Earth's surface as a means of evaluating the state of key geological questions in the mid-nineteenth century.

In Chapter 1, it was suggested that the first proponents of 'modern' earth science were trying to answer four questions.

1 What (and when) was the origin of the Earth?
2 What were the types of rock and what were their origins?
3 What were the origins of fossils?
4 What was the nature and origin of the Earth's topography?

By 1859, these questions had, to large measure, been disentangled partly as the expanding base of the scientific establishment led to greater specialization. The question of the Earth's origin and age was moving into the hands of physicists and (later) geophysicists, but with a minimum timeframe in millions, rather than thousands of years. Arguments about the origins of rocks had shifted well away from the Wernerian simplicities and, although there were still questions to be answered concerning the precise relations between intrusive and extrusive igneous material, and about the exact provenance of some superficial deposits, the notion of any universal connection between a rock stratum's lithological composition and its age was largely defunct. In its place was a growing realization that it was the *fossil* contents of rocks which could be used to provide temporal correlations, since there were distinctive suites of organisms found in clear stratigraphical relations. If a non-fossiliferous basalt layer was found sandwiched between fossiliferous beds A (below) and B (above), then the basalt was evidently younger than A and older than B (other things being equal). Why there was a general displacement of fossil flora and fauna over the extent of the fossiliferous geological record remained a matter of dispute and was, of course, one of the key answers ultimately provided by Darwin's views in *The Origin*.

It was the fourth question which, by 1859, was moving into prominence as the precursors of modern geomorphologists grappled with the evidence for the nature and relative efficacy of the two great classes of 'causes now in operation': the endogenetic forces of volcanoes and earthquakes, creating the broad outlines of global topography; and the exogenetic agencies of liquid water, ice and wind, responsible for the detailed sculpture of the surface. Even Lyell accepted that the balance between the different agencies had varied over time and place; but Agassiz's wholesale vision of global refrigeration and thawing was still only partly vindicated and its crucial concomitants in terms of eustatic (sea level) and isotatic (land level) fluctuation were just beginning to be appreciated.

Evidently, the powers one could ascribe to endogenetic and exogenetic causes now in operation hinged in real measure upon the antiquity one was prepared to accept for the Earth's origin. Similarly, the interpretation of changing fossil flora and fauna was associated with one's acceptance or rejection of ideas of progressive or non-progressive climate changes. And, for those of strictly Lyellian persuasion, all wholesale climate shifts that the geological record revealed – from the global deserts of the Triassic to the warm, shallow seas of the Cretaceous for instance – were only explicable in terms of essentially chance shifts in global topography. (Within our present paradigm of plate tectonics, that view in many ways reappears as a sensible one.)

As the Nineteenth century proceeded, there was not merely an expansion and professionalization and hence specialization of the earth science community, but also a huge expansion in global scientific travel. As we noted in Chapter 3, Charles Lyell was an indefatigable observer of European and eastern North American sites. Perhaps more important were the growing number of 'official' travellers. In Europe, these should be traced to the hugely influential missions of Bougainville (1766–69) and Cook (1769–1779) which opened the Pacific and its islands to increasingly scientific scrutiny. The tradition of naval vessels having a more or less qualified naturalist aboard (usually, though not inevitably, one of the surgeons) was most famously demonstrated by Darwin's participation in the *Beagle* voyage (1831–36) to be discussed below. But there were also the experiences of J.B. Jukes on HMS *Fly* on the Great Barrier Reef (1842–46) and of T.H. Huxley on HMS *Rattlesnake* (1846–1850). The success of the *Beagle* in scientific terms led to the U. S. Government mounting a massive 'Exploring Expedition' which spent five years (1838–1843) largely in the Pacific, providing James Dwight Dana (1813–1895) with the basis for his geomorphological ideas. This survey was succeeded by a procession of land-based, more or less official surveys of the American South West (*c.* 1855–1880) with important consequences which will be outlined in Chapter 6. (There was also the late and rather bizarre Harriman Expedition to Alaska and Siberia in 1899, in which G.K. Gilbert took part: Goetzmann and Sloan, 1982.)

It is probably always true that familiarity with particular landscape types will tend to colour the analogies made with new topographies and, hence, the preferred explanations concerning the balance of processes, past and present. It is extremely interesting to see Darwin's reactions to both the familiar and, of course, the unfamiliar.

DARWIN AND THE VOYAGE OF THE 'BEAGLE', 1831–36

Charles Robert Darwin (1809–1882) was, as is well known, only partly 'qualified' in any modern sense, to undertake the rôle of natural philosopher on HMS *Beagle*. His limited exposure to Wernerian geology, as delivered by Robert Jameson at Edinburgh University, had failed to excite (but see also Repchek, 2003); and it was purely fortuitous that a brief course of field geology in North Wales with the redoubtable Professor Adam Sedgwick directly preceded the invitation to join Captain Robert Fitzroy as both gentleman companion and naturalist and provided the 22-year-old with some vital practical skills. His grasp of botany and more especially zoology was rather more advanced, since he had been provided with excellent mentors at both Edinburgh and Cambridge. This is fairly general knowledge (see Browne, 1995; Keynes, 2002).

However, what is – or should be – quite staggering is how far the tyro 'philosopher' built on this limited practical background and – crucially – the volumes of Lyell's *Principles* (the first of which was a gift from Fitzroy) to make extraordinarily good sense of the multitudinous aspects of the natural world which passed before his eyes in the five years of the voyage. All those of us who have done independent fieldwork for the first time must remember the awful sinking feeling that comes with the realization that reality and the textbook differ. Even as supposed professionals, the bafflement by the new and unexpected leads – in my experience – to the frequent admission that 'I don't know what it is, but there's a lot of it about . . .'

When one contemplates (Figure 5.1) the geographical extent of *The Beagle's* voyage, encompassing tropical and temperate rainforests, hot and cold deserts, volcanic islands, coral reefs and atolls, the heavily glaciated landscapes of southern Chile, the grassy plains of Argentina, Australia and South Africa and the modern

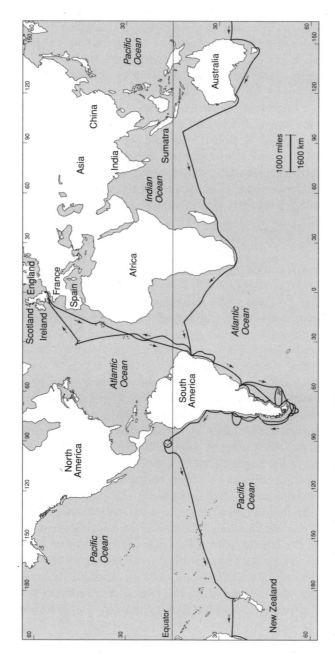

Figure 5.1 *Outline of the Beagle voyage, 1831–6.*

volcanoes and tectonic activity of Chile, it is a staggeringly large cross-section of the Earth's physiography, far greater than that observed even by the mighty Humboldt, by Pallas or by Lyell. Moreover, Darwin did not merely observe from the shoreline: he was (being often and hideously seasick) delighted to take off into the varied hinterlands of South America in particular. There was also the very significant journey with Fitzroy and the ship's boats up the great Santa Cruz valley of southern Argentina. By the time Darwin disembarked in October 1836, he was almost certainly the best-travelled natural scientist in the world.

More important, he proved, almost from the outset, that he had that rare genius which recognizes the crucial mismatches between text and nature (Dates from his *Journal of Researches*: 1860b, reprint of 1845, 2nd edn.) A key example comes in his account of the Santa Cruz expedition (April–May, 1834). This bright blue, fast-flowing river occupies a tiny fraction of a huge east–west valley, extending from the Atlantic in what looks like a straight line heading through the Andes to the Pacific (Figure 5.2). Both Fitzroy and Darwin were convinced that the valley was an ancient strait, upheaved along with the rest of Patagonia, but previously analogous to the Magellan Straits or the Beagle Channel. In fact, had the

Figure 5.2 *The Santa Cruz valley. (A) At Piedrabueno, near the estuary. (B and C) The middle valley. (D) The Perito Moreno glacier calving into Lago Argentino.*

(A)

(B)

Figure 5.2 *(cont'd)*

(C)

(D)

Figure 5.2 *(cont'd)*

expedition been able to continue ten or so kilometres west of the point at which they gave up, they would have seen before them, first, an enormous lake (Lago Argentino, discovered by Europeans 20 years later) and, at its head, massive east-flowing glaciers. The Santa Cruz valley is, in fact, now considered largely a product of glacial and fluvio-glacial activities. In 1834, however, Agassiz's Land Ice Theory was unborn: the explanation of wholesale extended glaciation was not available. But what is important is Darwin's clear recognition that the Santa Cruz is *not* a 'usual' fluvial valley. He checks, in particular, the way in which basalt from the multiple lava flows in the upper valley is transported by the river and concludes that this transport is so inefficient that the modern stream cannot possibly have carved the enormous trough in which it meanders. And he was absolutely right!

Overall, in fact, Darwin has what it would be easy to claim is a simple Lyellian preference for the power of marine action over fluvial. But we should be cautious. Darwin deliberately rejected Lyell's view on the origin of coral atolls (that they grow upwards from volcanic craters) and replaced it with his own theory which linked the suite of reef forms and atolls to zones of subsidence and elevation (which remains a basic tenet even now). So it is not good enough to think: 'Lyell said and so Darwin believed . . .' What we need to do is to think *what* Darwin saw and experienced.

First, he was perfectly clear about the power of fluvial erosion, especially on the various volcanic islands he visited. His account of the expedition he undertook into the interior of Tahiti (17–20 November 1835) makes plain his acceptance of the rôle of running water in the creation of the V-shaped valleys and intervening knife-edge ridges encountered (Figure 5.3A).

Second, he was clear that there was extraordinarily good evidence – including the Concepción earthquake of Chile which he experienced in March 1835 – for the general uplift of all of southern South America (see chapter 1 of his *Geological Observations on South America*, 1846). The general tendency for the wide, sloping valleys to have a surface layer of shingle, often including examples of seashells whose descendants were still to be found growing along the coast, appeared to reinforce the view that these valleys were once open estuaries – such as the Rio Plata – which had been quietly upheaved, allowing the seawater simply to drain away. In northern Chile, a whole sequence of such valleys extends northwards from Coquimbo into the Atacama, one which (the Quebrada Despoblada, Figure 5.3B, east of Copiapó) Darwin followed into the mountains (26 June to 1 July 1835). At that time, there was no stream nor even a dry watercourse in the valley; shingle was banked across the mouths of side valleys. Darwin could not be clearer about his conclusion: 'The sides of the crumbling mountains were furrowed by scarcely any ravines; and the bottom of the main valley, filled with shingle, was smooth and nearly level. No considerable torrent could ever have flowed down this bed of shingle; for if it had, a great cliff-bounded channel . . . would assuredly have been formed. I feel little doubt that this valley . . . [was] left in the state we now see by the waves of the sea as the land slowly rose' (1860b, pp. 427–8). Nowadays, however, there is a perfectly well-marked albeit generally dry stream

(A)

Figure 5.3 *Darwin's valleys. (A) V-shaped fluvial forms, Tahiti. (B) Broad, shingle-floored: Quebrada Despoblada (now Q. Paipote) near Copiapó, Chile. (C) Gap between islands, Chonos archipelago, Chile. (D) Tidal 'valley' network.*

channel (fed largely by snowmelt from the Andes) incising the great shingle fan. Any of the US explorers whom we shall meet in Chapter 6 would have had no thought for a marine rather than a fluvial origin for the equivalent tributaries of, say, Death Valley . . .

But third and most important, five years of extensive experience of the power of the sea, from both the fragile confines of the *Beagle* and from shorelines made it almost inevitable that Darwin would view the marine agency as by far the most effective of exogenetic forces. The currents and the storms which afflict the channels of Patagonia and Tierra del Fuego were notorious. The tidal range at the mouth of the Santa Cruz is more than 12 m. The meetings with icebergs laden

(B)

(C)

Figure 5.3 *(cont'd)*

with boulders showed how easy it would be for great erratics to be driven on-shore and dumped (cf. 1860b, p. 299). The passes between islands in the Chonos archipelago (Figure 5.3C) could easily be imagined as upheaved into the U-shaped valleys of the mainland, and, overall, the pattern of broad, flat-floored valleys with abrupt semi-circular heads occurred not only on many beaches (Figure 5.3D) but, it seemed, in many coastal locations. The most impressive of these last sites was in the Blue Mountains behind Sydney.

(D)

Figure 5.3 *(cont'd)*

'The country here is elevated two thousand eight hundred feet above the sea. Following down a little valley and its tiny rill of water, an immense gulf unexpectedly opens . . . at the depth of perhaps fifteen hundred feet. Walking on a few yards, one stands on the brink of a vast precipice, and below one sees a grand bay or gulf, for I know not what other name to give it . . . If we imagine a winding harbour, with its deep water surrounded by bold cliff-like shores to be laid dry . . . we should then have the appearance and structure which is here exhibited' (17 January 1836: 1860b, p. 523).

As Figure 5.4 shows, the Blue Mountain valleys are, indeed, spectacular, but modern studies (Bishop and Cowell, 1997) regard them as nevertheless due to fluvial action. The significance of the Blue Mountain vision in particular seems to have been responsible for Darwin's automatic comparison of the view of the North Downs and the Wealden valleys with the apparently incontrovertibly marine products of New South Wales. This is clear from an exchange of letters with Joseph Beate Jukes (1811–1869) in 1862 (Burkhardt *et al.*, 1997, pp. 219–221 and 227–9). Jukes, who together with Andrew Crombie Ramsay (1814–91) was to emerge as championing fluvial Denudation, had been perfectly happy with Darwin's estimated timescale for the Wealden denudation (see Darwin's letter to Asa Gray, 3 April 1860: Burkhardt *et al.*, 1993, p. 141) but he now takes up the cudgels with respect to the agent of erosion. Thus:

(A)

(B)

Figure 5.4 *The Blue Mountain valleys.*

'I suspect that the Chalk was bared of the Tertiary rocks by marine denudation as the rock rose above the Sea, that brooks commenced to run down the chalk slopes along the courses of those which now cut ravines through the Chalk escarpments, and that those ravines have been worn by those brooks continually cutting deeper than the ground inside . . . and the hills and valleys inside worn by the *rain* only and the *weather*. Your 300,000,000 of years is not nearly enough for the denudation of the Weald by this process' (25 May 1862; Burkhardt *et al.*, 1997, p. 210).

There is a missing letter from Darwin arguing for the efficacy of marine action in general. In his reply of 30 May 1862, Jukes sets out firmly the different modes of operation of marine and fluvial action, and then turns to what Darwin evidently considered the crucial case of the Blue Mountains:

'As to the valleys you mention in Australia I looked at some of them and have often considered your notions respecting them . . . Look at Sydney Harbour with its numerous arms all successively joining and going out by one narrow entrance. – I do not see how the Sea can possibly have made those ravines.

Look at the slope of the ascent of the Blue Mountains above the Hawkesbury [River] it is all gullied by similar ravines successively meeting, often emptying their brooks into one another through the narrowest and most precipitous gorges. I cannot but think they have all been cut by brooks when the land was higher and Sydney harbour above sea level . . .

But if so then those huge circular valleys you describe so well must nevertheless have been formed also on dry land however inconceivable it may be. – If the others are furrows worn in a sheet of sandstone those are only bigger holes worn perhaps by the union of some of the furrows. It is however quite possible that the erosion may have been helped a bit by the sea after the land had been down again and during its last rise –' (Burkhardt *et al.*, 1997, pp. 228–9).

It was, in fact, not until the 5th edition of *The Origin* (1869) that Darwin accepted that the sea must give way – in terms of topographic sculpture – to rain and rivers. As a retraction, it is both generous and wholesale, in my view:

'We have, however, recently learnt from the observations of Ramsay, in the van of excellent observers, of Jukes, Geikie, Croll, and others, that subaerial degradation is a much more important agency than coast-action, or the power of the waves. The whole surface of the land is exposed to the chemical action of the air and of the rain-water with its dissolved carbonic acid, and in colder countries to frost; the disintegrated matter is carried down even gentle slopes during heavy rain, and to a greater extent than might be supposed, especially in arid districts, by the wind; it is then transported by the streams and rivers, which when rapid deepen their channels and triturate the fragments . . .

Messrs. Ramsay and Whitaker have shown, and the observation is a most striking one, that the great lines of escarpment in the Wealden district and those ranging across England, which formerly were looked at by every one as ancient sea-coasts, cannot have been thus formed, for each line is composed of one and the same formation, whilst our present sea-cliffs are everywhere formed by the intersection of various formations. This being the case, we are compelled to admit that the escarpments owe their origin in chief part to the rocks of which they are composed having resisted subaerial denudation better than the surrounding surface; this surface consequently has been gradually lowered, with the lines of harder rock projecting. Nothing impresses the mind with the vast duration of time, according to our ideas of time, more forcibly than the conviction thus gained that subaerial agencies, which apparently have so little power, and which seem to work so slowly, have produced such great results' (1869, pp. 349–50).

Hutton and Playfair would certainly have agreed: I doubt Lyell did. (For further details on the chequered history of The Wealden Denudation, see Beckinsale and Chorley, 1991, pp. 285–298.)

DARWIN'S GEOMORPHOLOGY

It would evidently be absurd to claim Darwin as predominantly an earth scientist, let alone a geomorphologist. Yet it is too easily forgotten that it is in these fields that he first made his major academic mark. As Judd noted 'It is not too much to say that, had Darwin not been a geologist, the "Origin of Species" could never have been written by him. . . . I believe that the verdict of the historians of science will be that if Darwin had not taken a foremost place among the biologists of this country, his position as a geologist would have been an almost equally command-ing one' (Judd, 1890, pp. 270–271). There are – besides the *Beagle* diary or *Journal of Researches* (1839a) – four major works on geology and physiography as well as a number of early journal articles on which Judd's judgement is based. They cover between them an enormous sweep of topographic insight. The books are:

1842   *The Structure and Distribution of Coral-reefs*
1844   *Geological Observations on Volcanic Islands*
1846   *Geological Observations on South America*
1881   *The Formation of Vegetable Mould through the Action of Worms, with Observations on their Habits*

Among the purely 'geological papers' was one of Darwin's rare, total blunders: *Observations on the Parallel Roads of Glen Roy* (1839b). Here, just as Agassiz was preparing to launch his land ice vision, Darwin decided to attribute the mysterious parallel lineations on the side of Glen Roy to marine abrasion. As he said 'It is admitted by every one, that no other cause, except water acting for some period on the steep side of the mountains, could have traced these (terrace) lines over an extensive district' (1839b, pp. 39–40). After exposure to Agassiz's revelations and the detailed investigations by T.F. Jamieson (in 1861, published 1863) demonstrating pretty conclusively that the 'roads' were the former shorelines of freshwater, glacially dammed lakes, Darwin was quite evidently mortified by his lack of perception 'I am smashed to atoms about Glen Roy. My paper was one long gigantic blunder from beginning to end' (6 September, 1861: Burkhardt *et al.*, 1994, pp. 256–7). Despite a few subsequent twinges of doubt, Darwin acted as an extremely positive referee for the publication of Jamieson's findings (see Burkhardt *et al.*, 1999, pp. 93–4; and Burkhardt *et al.* 1994, Appendix IX, pp. 429–59).

Although in some ways more receptive than Lyell to Agassiz's land ice vision, there can be no doubt that Darwin found it difficult to dismiss the rôle of icebergs in the rafting of erratic materials inland (cf. his letter to G.C. Walich, 12 December, 1860: 'Do you not think you are rather bold in inferring that the basaltic pebbles were rounded at such great depths? Are you sure that they were not dropped by icebergs either recently or at the close of the Glacial period?' (Burckhardt *et al.*, 1993, p. 526).

The overtly glaciated landscapes he had seen were, characteristically, those of the steep western coast of Chile (see the discussion in chapter X of *The Beagle Voyage*, 1860b), whose valley glaciers in some cases reach to sea level. Inland – either on his traverses of the Andes or in his excursions on the Patagonian plains – there was little or no direct visible link between land ice and landscape. One exception which gave him some pause was the fact that, as they ascended the Santa Cruz valley, the gravel deposits thickened inland. This was curious, for an ostensibly marine formation (which should have been thickest near the coast). But the gravel was *rounded* (it is undoubtedly fluvio-glacial and linked to great glacial melt floods), not the angular material to be found at glacier snouts, so that it was evidently waterlain . . .

However, what Darwin was better at than Lyell (and many others among us!) was admitting the force of counter evidence and changing his mind. He did, with respect to the dominant rôles of both fluvial and glacial action.

He was also forced to retreat over his excessive estimate of the age of the Wealden denudation (now considered to have taken about 60 million years: Jones, 1981). But, even hard-pressed by the physicists, he could not give way over the extensive timescale that he knew was the only possible explanation for his views on both landscapes and organisms. This need for 'deep time' and the significance of the action of minute processes was inherent in his work on both coral reefs and vegetable mould.

In the case of coral reefs, 'the polypifers which construct these vast works' (Judd, 1890, p. 11) are minute, yet the volumes of the reefs they construct are immense, in both horizontal and – if Darwin was correct – vertical extent. Moreover, these polypifers were able to outperform both the endogenetic agencies and the attack of the sea. As Darwin concludes 'Reflecting how powerful an agent with respect to denudation, and consequently to the nature and thickness of the deposits in accumulation, the sea must ever be, when acting for prolonged periods on the land, during either its slow emergence or subsidence . . . I may be permitted to hope, that the conclusions derived from the study of coral-formations, originally attempted merely to explain their peculiar forms, may be thought worthy of the attention of geologists' (Judd, 1890, pp. 109–10). At the end of his life, they were indeed still attracting the attention – and the arguments – of geologists. Darwin was still convinced he was right, but recognized that there was a simple way to resolve the debate; about a year before his death, he wrote:

'If I am wrong, the sooner I am knocked on the head and annihilated so much the better . . . I wish some doubly rich millionaire would take it into his head to have borings made in some of the Pacific and Indian atolls, and bring home cores for slicing from a depth of 500 or 600 feet' (quoted in Judd, 1890, p. 9). It was the USA Government who finally performed this function, after atom tests in the 1950s. Darwin's confidence was vindicated (Beckinsale and Chorley, 1991, p. 95).

This belief in the power of the lowly dominated Darwin's last work (1881) on the rôle of earthworms in denudation. It had an interesting origin: his father-in-law Josiah Wedgwood II had shown him how rubble strewn on meadows had vanished under a layer of loam in a matter of a few years (1881, pp. 132–9). Struck by this, Darwin prepared a paper on the phenomenon for the Geological Society in 1837, which was not well received. But he did not give up and, for over 40 years, the Downe postman brought not only pillboxes of barnacles (see Stott, 2003) but (rather soggy, one imagines!) packets of earthworm casts from the four corners of the world. Two parts of a field at Downe were spread with either chalk (in December 1842) or cinders (in 1842–3) and excavated from beneath the earthworm layer in November 1871. After careful observation, Darwin's conclusion was:

'Where the land is quite level and is covered with herbage, and where the climate is humid so that much dust cannot be blown away, it appears at first sight impossible that there should be any appreciable amount of sub-aerial denudation; but worm-castings are blown, especially whilst moist and viscid, in one uniform direction by the prevalent winds which are accompanied by rain.
    The removal of worm-castings by the above means leads to results which are far from insignificant . . . It was found by measurements and calculations that on a surface with a mean inclination of 9°26′, 2.4 cubic inches of earth which had been ejected by worms crossed, in the course of a year, a horizontal line one yard in length; so that 240 cubic inches would cross a line 100 yards in length . . . Thus a considerable weight of earth is continually moving down each side of every valley, and will in time reach its bed. Finally this earth will be transported by the streams . . . into the ocean. . . . So that, if a small fraction of the layer of fine earth, two tenths of an inch in thickness, which is annually brought to the surface by worms, is carried away, a great result cannot fail to be produced within a period which no geologist considers extremely long' (1881, pp. 309–11).

It was to be almost 80 years before anyone else took up the prospect of worms as a geomorphological agent (Kirkby, 1967); but we can now see a wholesale interest in the rôle of organisms (and especially micro-organisms) in denudation (cf. Viles, 1988; Kennedy, 2000).

CONCLUSION

Darwin was a genius, probably the only student of landforms to deserve to be labelled a genius. The range and – in general – the perspicacity of his observations of the forces now in operation at the Earth's surface were, as I hope to have demonstrated, profound.

He lived to see the development of a far better balanced vision of the interplay of glacial, fluvial and marine agencies than had been put forward in Volume I of Lyell's *Principles* which was his initial *vade mecum* on the great *Beagle* voyage. One of the clearest visions was provided by his fervent evolutionary champion, Thomas Henry Huxley (1825–1895) in the first edition of his work *Physiography*, published in 1877 (see Stoddart, 1975). Huxley's work can fairly claim to be the first fluvialist textbook (albeit retaining a rôle for marine planation: Chorley *et al.*, 1964, pp. 595–6) although, as Stoddart has shown, the geomorphologic content was progressively removed in later editions. But the vindication of Hutton and Playfair's vision of a predominant rôle of rain and rivers in the sculpture of the land surface came, not from Europe, but from the American Southwest and the great canyons of the Colorado Plateau. The Grand Canyon was enough to convince anyone of the power of running water . . .

Chapter 6

Inventing a Fluvial Landscape: Powell, Gilbert and the Western Explorations

'And what a world of grandeur is spread before us! Below is the canyon through which the Colorado runs . . . From the northwest comes the Green in a narrow winding gorge. From the northeast comes the Grand, through a canyon that seems bottomless . . . Away in the west are lines of cliffs and ledges of rock . . . ledges from which the gods might quarry mountains; and . . . cliffs where the soaring eagle is lost to view ere he reaches the summit' (J.W. Powell, July 19, 1869; in Powell, 1895, p. 212).

INTRODUCTION

Both James Hutton and John Playfair had evidently possessed a vision of the geological cycle (see Chapter 2) in which the dominant denudational rôle, over the whole range of time and space, belonged to rain and rivers (see Repchek, 2003). As Playfair put it simply: '[O]n our continents, there is no spot on which a river may not formerly have run' (1802, p. 352).

However, as we have seen, the traces of rivers were, in much of the Northern Hemisphere, severely confused by the overprint of recent intermittent glacial activity and of concomitantly oscillating sea levels. By the mid-1860s, British and American workers, including Darwin, were almost at the point of being able to disentangle the various agencies: Ramsay's 1862 paper demonstrating the glacial origin of deep lakes such as Lake Geneva was – as we have seen in Chapter 3 – extremely important. Similarly significant was Jukes's account, also in 1862, of the evolution of the valleys of southern Ireland, which provided a non-marine explanation for the way in which small rivers came to flow straight through ridges of high ground. This puzzling phenomenon also occurs in close proximity to Darwin's house at Downe in Kent and the way in which the puny River Darenth punches through the North Downs was only satisfactorily ascribed to a similar combination of uplift and fluvial incision, by Ramsay, in 1863. The next years saw the application to topography of the crucial theory of glacial isostasy (by T.F. Jamieson) and a renewed emphasis on glacial eustasy (by S.V. Wood). However, it was still thought that the Ice Age had been only a single episode. Also, the vexed question of absolute dating was nowhere near being solved. Nevertheless, by sheer slogging – if you like, by a collective operation of the Method of Multiple

Figure 6.1 *The Grand Canyon: from Huxley's* Physiography, *2nd edn.*

Working Hypotheses and following Sherlock Holmes' dictum that 'when you have eliminated the impossible, whatever is left, however improbable, must be the truth' (Conan Doyle, 1890), the basic situation was coming clear. Rain and rivers are the bottom line of earth sculpture: wind, ice and the sea are players only in specific geographical and temporal settings. However powerful, they are not and probably never have been, the dominant agencies of denudation for the whole globe.

It had taken nearly 100 years to create what we would consider such a balanced view of earth surface processes culminating in Huxley's *Physiography: an Introduction to the Study of Nature* (1877) which began with a consideration of the origin of the Thames and its basin (see Stoddart, 1975). Two very significant inclusions were J Tyndall's observations (1860) on Alpine glaciers and, of direct relevance here, J.W. Powell's 1875 *Exploration of the Colorado River of the West* (see Figure 6.1).

POWELL AND THE COLORADO PLATEAUS

Until the end of the Civil War, virtually all scientific study of North American landscapes was confined to the eastern and northeastern regions. In the latter instance, the problems posed by an imperfectly recognized glacial inheritance were as substantial as those in north and west Europe (see Tinkler, 1985).

One important addition to topographical knowledge came from James Dwight Dana (1813–1895) who, profoundly influenced by his participation in the massive U.S. Government's Exploring Expedition (1838–1843) brought out the first edition of his *Manual of Geology* in 1863. This is, to be blunt, a curious work, in that it has a fundamentally teleological vision (see Kennedy, 1992) in which an 'infinite mind has guided all events towards the great end – a world for mind, – [and] the earth has, under this guidance and appointed law, passed through a regular course of history and growth' (1869, 2nd edn, p. 1).

As Darwin says, frankly disparagingly, in an earlier letter to Asa Gray, apropos Dana 'I believe, poor fellow, that he believes in 1st Chapter of Genesis, so great allowances must be made for him' (1857, Burkhardt and Smith, 1990, p. 516). In many ways, with his vision of successive epochs of earth history closed by cataclysmic revolutions, Dana harks back to Buffon or Cuvier. (Or, even, to Robert Jameson, whose insistence on an overall Wernerian plan was similarly punctuated by very acute, 'modern' observations of the role of modern processes, including glaciation: see Davies, 1968, pp. 267–8). For instance: 'Many examples are on record of gorges hundreds of feet deep cut out of the solid rock by two or three centuries only of work' (Dana, 1869, p. 639). There is in fact, as with Lyell, a tendency to overstate the significance of the occasional 'debacle' and, hence to speed up fluvial denudation: Dana considers (1869, p. 570), for example, that all the globe's major river systems are Post-tertiary (i.e. Quaternary: which would imply, we would think, that they were a mere 2–3 million years old). Although this may, indeed, be the case for the rivers, the *valleys* are almost always substantially older. Both rivers and valleys have often experienced major modifications as varying climates and tectonics have impinged upon them. (See also Ollier, 1991.)

The evidence for wholesale fluvial activity which in Dana's case had undoubtedly originated in his views of the heavily dissected volcanic islands of the Pacific, had dramatically increased since 1857, when J.S. Newberry (1822–92) accompanied J.C. Ives in an expedition to the lower Colorado River (Chorley *et al.*, 1964, p. 501). F.W. Von Egloffstein, the German topographer who was in the party, presented the world with the first illustration of the Grand Canyon. As Chorley *et al.* note (p. 503), this is a distinctly 'Gothic' vision (one expects to see Count Dracula hovering around a pinnacle) (cf. Figure 6.2). More crucially, it is of a deep, narrow

Figure 6.2 *A canyon. (Bonney, 1896.)*

slot. One of the key difficulties put forward apropos the fluvial origin of valleys – and Darwin, I think, had this problem – was how one went from narrow gorges or steep-sided V shapes to the open, shallow cross-sections of (say) the Vale of Holmesdale at the foot of the North Downs. Despite the evidence from Europe provided by George Poulett Scrope (1797–1876), and the frankly intemperate fulminations by George Greenwood (1799–1875) in his 1857 *Rain and Rivers: or Hutton and Playfair versus Lyell and all comers*, the widespread production of broad *coastal* platforms seemed to urge the predominance of waves in creating wide, gently sloping surfaces. The significance of the Colorado Canyon lay not simply in its classic fluvial shape, but in its intimate linkage with the great suites of plateaus and cliffs which surround it (Figure 6.3).

It was the work of John Wesley Powell (1834–1902: see Worster, 2001), which not only hammered home the connection between vertical and horizontal fluvial denudation in the Colorado region, but also provided some basic rules (which have proved to be durable) for the operation of the processes concerned.

Powell was an extraordinary figure (Chorley *et al.*, 1964, chapters 25 and 27). After an unfocused early career as teacher and traveller, he enlisted in the Federal Army at the outset of the Civil War in 1861, but, within the year, he lost his right arm at the battle of Shiloh. Despite this, he continued in active service (until January 1865), being promoted to Major. On resigning his commission, he obtained a post as Professor of Geology in Bloomington, Illinois, and in 1867 set out on the first of five expeditions in the Southwest. The most important of these were the third, in 1869, which culminated in the first traverse of the Grand Canyon by Europeans, and the fourth (in 1871–2). These also produced the enormously influential *Exploration of the Colorado River of the West* (1875), *The Report on the Geology of the Eastern Plateau of the Uinta Mountains* (1876) and their successor *Canyons of the Colorado* (1895), which brought the landscapes of the area to general attention. The illustrations (cf. Figure 6.3) show quite incontrovertible evidence that vertical incision and horizontal retreat of plateau edges have gone hand in hand. The relative absence of vegetation cover made the whole situation crystal clear: as Powell says: 'All about me are interesting geological records. The book is open and I can read as I run' (and 'run' he most certainly did: it takes a perpetual mental effort to recall that the author who talks blandly of climbing up to the top of the canyon wall – after a hair-raising day's boat trip – was doing this scrambling, literally, single-handed).

But just as Darwin went beyond a traveller's account in his development of a working hypothesis to explain the variety of coral reefs, so a fundamental analysis of geological (and geomorphological processes) emerges similarly from Powell's accounts.

First, he uses the unconformities visible in the canyon walls (recall Hutton's example, Figure 2.2) plus the stepped sequence of cliff-bounded plateaus to invent the concept of *base level*. 'It should be observed that no valley can be eroded below the level of the principal stream, which carries away the products of its surface degradation . . .' (1875, p. 163). 'We may consider the level of the sea to be a grand base level, below which the dry lands cannot be eroded; but we may also have, for local and temporary purposes, other base levels of erosion' (pp. 203–4).

Second, Powell creates a firm *framework* which links the development of river valleys to the uplift of the land surface. Unlike Dana's frankly descriptive (and boring) catalogue of 'the reliefs or surface-forms of the continents' (1869, pp. 23

Figure 6.3 *The Grand Canyon, showing horizontal as well as vertical fluvial erosion. (Bonney, 1896.)*

et seq.), Powell puts forward the fundamental genetic classification which links streams and tectonics: antecedent (there is uplift across the valley axis and the river continues to erode, successfully, without deflection of its course, as, most spectacularly, in the Himalayas); superimposed (a river which began flowing on an elevated stratum, continues to cut down into folded, lower rocks: with the removal of the higher layer, the river is left flowing through high ground – the Darenth in Kent is an example); and consequent (where the stream flows accordantly down the dip of the rocks, as on the dip slope of the North Downs). These ideas were (possibly) not original to Powell, as both Jukes in his Irish study (1862) and F.V. Hayden (1862) on the upper Missouri had arrived at some of the same concepts, but it was certainly Powell who systematized the description.

Finally, Powell devoted attention to the way in which the forces of subaerial denudation interact with the nature of the surfaces on which they operate. 'The lesser or greater rapidity of erosion' [by running water] 'depends chiefly on three conditions: first, elevation above the base level of erosion; second, the induration of the rocks; and, third, the amount of rain fall' (1876, p. 34). Here Powell explicitly identifies two factors – relative relief and precipitation – which provide the potential and the kinetic energy for erosional processes and a third – the relative resistance of the strata – which acts to limit the success of such erosion. Powell is particularly interested in the manner in which the relationship between 'rainfall' and the 'rapidity of erosion' is dependent upon the extent of vegetation cover:

> 'And yet the conditions necessary to great erosion in the valley of the Colorado are not found to exceed those of many other regions. In fact, the aridity of the climate is such that this may be considered a region of lesser, rather than greater, erosion. We may suppose that, had this country been favoured with an amount of rain-fall similar to the Appalachian country, and many other districts on the surface of the earth, the base level of erosion in the entire area would have been the level of the sea; and, under such circumstances, although the erosion would have been much greater than we now find, the evidence of erosion would have been more or less obliterated' (1875, p. 209).

These ideas have remained crucial elements in geomorphological thinking. Powell's three, rather precisely defined controls, were to resurface later in the century as the Davisian trio of 'stage, structure and process' (or, more usually, 'Structure, process and stage': see Chapter 7); and 80 years or more on would see Langbein and Schumm (1958) asserting that the maximum rate of fluvial erosion coincides with semi-arid environments (sufficient kinetic energy but inadequate surface protection). (See Chapter 8.)

After the expeditions, Powell became desk-bound as the second Director of the U. S. Geological Survey and, as far as our present story goes, rather fades from view. However, one of his key legacies to the wholesale scientific onslaught on the topography of the West lay in Powell's employment and encouragement of Grove Karl Gilbert (1843–1918).

GILBERT'S LANDSCAPES

Relatively few of the earth scientists whom we have introduced in this account have retained any major stature in modern geomorphological thinking. A nod here and there to Hutton, Playfair, Agassiz and Darwin (cf. Huggett, 2003), but – by and large – the contributions of these investigators are seen as of little direct relevance to our modern concepts of earth surface processes and landforms however significant were their ideas (see Repchek, 2003, for example). Gilbert, on the other hand, is often regarded today in key respects as a precursor of the 'invention' of the mathematical and physical reductionist approach to geomorphology to be described in Chapter 8. In this sense, he is also cast as the St George figure who helped to slay the Davisian dragon (Chapter 7). This view of Gilbert, I maintain, is a true invention (see Kennedy, 1992) and represents the all-too-common attempt to trace a 'proper' pedigree for 'modern' positions. (I find this tendency especially marked in most of the essays in the commemorative volume edited by Yochelson, 1980.) The truth, I think, is more interesting.

Like Powell, Gilbert – after a standard general college education – tried his hand as a teacher before opting for life as a surveyor. He spent three years on the Western survey of G.W. Wheeler before joining Powell's team. Thereafter, his life oscillated between field (and laboratory) studies and a Washington desk, as he was prevailed upon to join the senior administration ranks of the U.S. Geological Survey under Powell's Directorship in 1881.

Although Gilbert published on a wide range of topics (see Kennedy, 1992, 1993; Sack, 1992a,b) there are five studies in particular which have led to his adoption as the key forerunner of the mid-twentieth century love affair with the physics and mathematics of earth surface processes. These are:

1877	*Report on the Geology of the Henry Mountains* (actually published 1879)
1885, 1890	*Contributions to the History of Lake Bonneville*
1909	*The Convexity of Hilltops*
1914	*The Transportation of Debris by Running Water*
1917	*Hydraulic Mining Debris of the Sierra Nevada*

In all of these, Gilbert's focus can be interpreted as that of the physical or experimental scientist (cf. Pyne (1980, p. 95) who claims Gilbert as a 'pure Newtonian') concerned to tease out the fundamental processes – and laws – which govern the movement of water and sediment both in fluvially eroded landscapes and along shorelines (as in the case of Lake Bonneville: see Hunt, 1980). In the last paper listed above he moves on to what was to become a great late twentieth-century obsession: the human impact (cf. Goudie, 1981 *et passim*): see Chapter 9.

There should be no doubt that Gilbert does, indeed, make major statements which fit this image. In the Henry Mountains paper (described by Chorley *et al.*, 1964, as his 'masterpiece'), most attention is devoted by modern workers to chapter 5, where Gilbert is actually describing the processes in the North Caineville

Mesa badlands (Figure 6.4) at the northern end of the great pile of intrusive igneous material (laccolites) which constitutes the Henrys. (As an aside, Gilbert was the first to name the laccolites and he chose to commemorate Newberry, Dana, Scrope, Geikie and Jukes amongst his contemporary geomorphologists with a laccolite apiece.)

In Caineville, where areas of soft sediments are exquisitely regularly dissected by running water, Gilbert developed several key visions of the way in which running water operated (much of which was published in an article in the *American Journal of Science* in 1876). A crucial concept, which has fascinated or bedevilled more than a century's worth of geomorphological thinking, was that of *grade* defined in the 1876 article and repeated in the Henry report as follows:

> 'In general, we may say that a stream tends to equalize its work in all parts of its course. Its power inheres in its fall and each foot of fall has the same power. When its work is to corrade' [i.e. erode mechanically] 'and the resistance is unequal, it concentrates its energy where the resistance is great, by crowding many feet of descent into a small space; and diffuses it, where the resistance is small, by using but a small fall in a long distance. When its work is to transport, the resistance is constant, the fall is evenly distributed by a uniform grade. When the work includes both transportation and corrasion, as is the usual case, its grades are somewhat unequal; and the inequality is greatest when the load is least' (1877, p. 113).

As Chorley *et al.* note (1964, p. 550) this passage has what was to prove an inherently confusing tendency to conflate a graded *condition* with an even *gradient*. In the Henry Mountains report Gilbert moves the concept forward to describe a state in which there is neither (net) erosion nor deposition along the channel. 'A fully loaded stream is on the verge between corrasion and deposition. It may wear the walls of its channel, but wear of one wall will be accompanied by an addition to the opposite wall' (1877, p. 111). Here the state is more significant than the gradient: a good example of the kind of condition Gilbert envisaged occurs in meanders, where erosion on the concave bank (often resulting in a river cliff) is effectively matched by deposition on the convex, slip-off slope.

Nevertheless, Gilbert undoubtedly paid great attention to the way in which the gradient of both slopes and steams governed the rate of their evolution. The Henry report produced a number of laws, which have *gradient* as a key element: the *law of uniform slopes* suggested that, the steeper an initial slope, the faster it will be eroded, thus ironing out any topographic differences (and creating the visually stunning regularity of the Caineville badlands). The *law of divides* – which applied equally to slopes and steams – explained their changing slope and therefore generally concave-up curvature by the relative volume of water flowing down them. The small volumes near the head led to limited denudation and, hence, steep slopes. These became progressively flatter as the volume of flow increased down valley. This created sharp crests between two concave surfaces: the crest would also move towards the less active side, which – in turn – would steepen that gradient and lead to a return to balance. Not all of the divides in the Utah badlands

(A)

(B)

Figure 6.4 *Gilbert's Caineville Mesa field areas. (A and B) Sharp-crested erosion of badlands. (C) Mancos Shale rounded divide.*

(C)

Figure 6.4   *(cont'd)*

were, however, sharp (recall Figure 6.4C) and it was not until 1909 that Gilbert proposed that it was the relatively diffuse process of surface soil creep which produced such convexities. (This in turn, led to a key concept in R.E. Horton's 1945 hydrophysical approach to fluvial processes: see Chapter 8.)

This version of a balance between opposing forces emerged in three other, extremely important, ideas developed by Gilbert.

First, looking at the range of hillslopes, Gilbert recognized that there was a continuum observable from bare surfaces to those totally covered by soil and vegetation. He ascribed the difference to the balance between the forces causing disintegration of the solid slope material – principally weathering – and the export of the comminuted products by the agents of transportation (1877, p. 119). A *weathering-limited* slope, in the limit, is a naked cliff face: any particles or chunks dislodged from the surface – by gravity, wind, water, ice or organisms – fall or roll clear and leave the face open to renewed attack. A *transport-limited* surface, on the other hand is, like Darwin's Kentish hillside (see above, p. 70), prone to extremely slow development as the products of weathering or erosion build up. In the limit, only biological agencies have the energy available to continue the development of these slopes (see Kennedy, 2000). For most hillsides, Gilbert saw the balance between weathering and erosion/transportation as determining the slope form. This, in turn, often depends upon the mechanical properties of the slope materials: hard, dense, massive rocks are more apt to resist weathering and form towering cliffs than (say) soft clays and shales: see Selby (1993) for a comprehensive modern discussion of slope processes.

Second, Gilbert's vision of balance was very extensively applied to the work of rivers: the amount ('capacity') of debris of a given grain size which can be carried ('the steam's competence'). Gilbert showed the latter to vary with the channel's gradient and the volume of the water flow (we now say 'stream power': cf. Knighton, 1998), and, crucially, the lithological composition of the bed and banks. It is this crucial balance between force and resistance to which Gilbert returned in a hugely important discussion in his 1914 blockbuster. Self-evidently, he thought, the way in which the power of the stream can be developed – and, hence, the nature of the channel processes and forms – will hinge upon the relative erosivity of the bed and bank materials. He identifies in the 1914 paper (p. 219) three distinct categories of channel (Figure 6.5A–C) and I would add the logical fourth (Figure 6.5D):

1 alluvial (mobile)
2 rock-walled
3 corrading (rock bound)
4 rock floored.

In my opinion, much of the process-based discussion of fluvial geomorphology in the mid-twentieth century was inherently flawed (cf. Dury, 1964–5) by a failure to grasp Gilbert's forceful recognition that the balance of force and resistance varies with channel boundary type. This led to an emphasis on the rôle of discharge alone as the determinant of channel form and change. As the channels investigated were almost exclusively alluvial, this seemed to make sense. However, work from Australia, in particular, has shown that rock walled or rockbound or rock floored channels do, indeed, respond differently. (See the range of studies in Miller and Gupta, 1999.)

Third, in his Lake Bonneville studies, Gilbert was directly concerned with the idea of balance, but in this case between the action of waves and of fluvial and slope processes.

What emerges, if you look at Gilbert purely in this light, is the picture of a classical physical scientist whose vision is of a landscape everywhere striving towards some ultimate equilibrium between force and resistance. If it were to

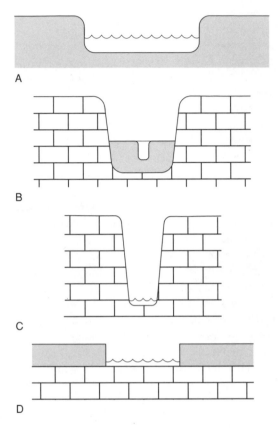

Figure 6.5 *Types of river channel, after Gilbert (1914) and author: (A) mobile, (B) rock bound, (C) rock walled, (D) rock floored.*

exist, such a landscape would be both timeless and unchanging, and ever changing: call to mind the ever-altering yet apparently never-changing state of a sandy beach.

To think of Gilbert in this way, however, as I have argued (Kennedy, 1992), is to very seriously underestimate the range and power of his approach to landforms and landscapes. It is part of the battle between the experimental and the observational (or historical) scientist which we have experienced since Kelvin (the experimentalist) took up the cudgels to trounce Darwin (the historical scientist) about the Earth's age. It is too easy to forget that Kelvin WAS WRONG. (As I have already stated, Burchfield's 1990 account gives one the clear impression that Kelvin's erroneous estimates of the Earth's age, impeccably based on unwittingly incomplete physical principles, which were – at worst – out by something like 4500 million years, were nevertheless far more creditable than Darwin's Wealden figures – only out, we would say, by 240 million . . .).

A similar conflict between the observational evidence of Wegener and the 'known' of geophysics (see Oreskes (1999) for a fascinating discussion) was, ultimately, resolved in Wegener's favour.

The world's landscapes all have historical legacies. It is still – and, in my view, always will be – impossible to believe the boast of one ultra-reductionist geomorphologist whom I heard declare 'Tell me the forces operating on a particle of silt and I will explain the evolution of the Himalaya mountains . . .'. Gilbert, I maintain knew this. Just as Darwin could vary his approach to problem-solving to suit the problem (from the map which is the key evidence for his coral reef theory to his experiments with the viability of seeds after immersion in salt water or his long-term experiments with worm activity), so, too, could Gilbert. A responsibly supported explanation of the Henry Mountains, as a large and complex mass of partially eroded intrusive igneous features termed laccolites, required far more than a recognition of the balance between force and resistance in the modern gully system of the adjacent Caineville Mesa. It required a careful analysis *both* of the probable physical forces at work controlling the emplacement of the laccolites *and* of the sequence of events leading to their present topography. Gilbert does precisely this in chapter 4 of the Henry report. The process-dominated studies of the late twentieth century (see Chapter 8) never quote it.

There are two other papers by Gilbert which I would commend to anyone wishing to see how he thought.

The first would be his experimental study of the Moon's craters (1893): he gives a fascinating account of his – largely successful – attempts to distinguish the topographic result of impact and of explosion. This is probably the first genuinely modern scientific attempt at non-terrestrial geomorphology (see Kennedy, 1993).

Second, there was his paper (1896) on what is now known as Meteor Crater, Arizona (then Coon Butte) as a focus for his exposition of 'scientific method'. This paper has many similarities to the later and greater Method of Multiple Working Hypotheses exposition by T.C. Chamberlin (1890: see Chapter 1). Alas – as with Darwin and the Glen Roy terraces – Gilbert reached what we now consider to be a false conclusion. But that matters less than the elegance with which he sets out his methodology.

'Phenomena are arranged in chains of necessary sequence. In such a chain each link is the necessary consequent of that which precedes, and the necessary antecedent of that which follows . . .

It is the province of research to discover the antecedents of phenomena. A phenomenon having been observed, or a group of phenomena established by empiric classification, the investigator invents an hypothesis in explanation. He then devises and applies a test of the validity of the hypothesis. If it survives the test, he proceeds at once to devise a second test'.

It is worth repeating the key passage (from Chapter 1) which states that in the testing of hypotheses lies the prime difference between the investigator and the theorist.

'The great investigator is primarily and pre-eminently a man who is rich in hypotheses. In the plenitude of his wealth he can spare the weaklings without regret . . . The man who can produce but one, cherishes and champions that one as his own, and is blind to its faults. With such man, the resting of alternative hypotheses is accomplished only through controversy' (Gilbert, 1886, pp. 286–7).

Gilbert was, it seems, an essentially modest man (see Chorley *et al.*, 1964, chapter 28) and yet the description of the great investigator clearly matches him to a T. The developing schools of geomorphology could and should have taken heed of Gilbert's views, as well as T.C. Chamberlin's later extension and formalization of the methodology. I fear that, here again, the late twentieth century's lip service to Gilbert's legacy seems to have largely gone without any substantial grasp of his advice.

I have said in the last chapter that Darwin was a genius some of whose brilliance, persistence and penetration constituted an important contribution to the study of landforms. If I were asked to name a 'pure' earth surface scientist who comes close to Darwin's breadth and intelligence of problem solving, it would be Gilbert. Quite when, why and how he developed the skill and insight which allowed him to succeed so brilliantly as both an experimental *and* a historical scientist is a mystery to me. Perhaps we are too ready to assume that all really important figures are 'all of a piece'. It is, however, ironic that the overt mantle of an ostensibly Darwinian view of landforms passed – not to Gilbert – but to his American contemporary, William Morris Davis, with far-reaching consequences.

Chapter 7

Inventing the Geographical Cycle and the Synthetic Genius of W.M. Davis

'His genius was of a particular sort. He created nothing – nothing, that is, except coherence and vitality; and his synthetic means of so doing was as simple and obvious as the manner in which Lazarus was raised from the dead' (Chorley *et al.*, 1964, p. 621, on William Morris Davis (1850–1934)).

INTRODUCTION

Pretty much any English-speaking earth scientist who graduated before 1970 can tell you something about the Harvard Professor who came to bestride the geomorphological world for the first half of the twentieth century. Some will still revere him. Some will execrate his influence. But the name of William Morris Davis will definitely not pass without reaction amongst such individuals. Some measure of the breadth and depth of his impact is given by the first three volumes of *The History of the Study of Landforms* (Chorley *et al.*, 1964; Chorley *et al.*, 1973; Beckinsale and Chorley, 1991). In the first, Davis and his remarkable invention 'The Geographical Cycle' are accorded only one chapter (the last). The second – all 874 pages of it – is subtitled 'The life and work of William Morris Davis'. The third (subtitled 'Historical and regional geomorphology 1890–1950') has one of its four Parts devoted explicitly to 'Davisian influences'. Although there were other important actors at work upon the increasingly professional and international geomorphological scene of the late nineteenth and first half of the twentieth century, it is evident that Davis warrants attention. Just why was his influence so profound?

THE IMPORTANCE OF DISCIPLES AND PUBLICATIONS

None of those whom I have identified so far as key contributors to the particular version of Earth history and sculpture to which I happen to subscribe – Hutton, Playfair, Lyell, Darwin, Powell and Gilbert – could genuinely be described as 'Founders of a School'. By that I would mean that they attracted pupils – not just admirers – who actively proselytized The Master's word and who, in turn, attracted pupils themselves, thus founding intellectual dynasties. One prerequisite for such a development is, it seems, tenure of a relevant academic position. In this respect, Hutton had no academic base; Playfair was a Professor at Edinburgh University, but of Natural Philosophy and Mathematics; Lyell held, briefly, a chair at the new King's College, London but gave lectures rather than running a research department (Wilson, 1972); the increasingly reclusive Darwin's influence spread by correspondence, rather than direct contact; Powell became an administrator and so did Gilbert. The major successful proselytizers we *have* met were Werner, secure in his post at the Saxon School of Mines in the latter part of the eighteenth century; and one of *his* disciples, Robert Jameson, ruling the academic roost at Edinburgh in the early nineteenth century. Louis Agassiz, at Harvard, was a later and dominating figure as Hartt's discussion (1870) of his Amazon expedition (1865–6) makes plain.

As our story moves towards the present, so it is vital to recognize that there are 'families' of academics, whose central tenets have been established during their periods as undergraduates, graduates or junior academics within the orbit of some key and often distinctly charismatic figure. In some instances – as with some families – the influence is entirely benign and apostles simply encouraged by example. In others – and Davis seems to have been a case in point – there develops a very rigid vision of 'the right way', which not merely demands methodological obedience from acolytes, but The Master may resort to punitive measures if there is the suspicion of mutiny in the ranks. Of course, again as in some families, it is not unknown for there to be revolt against overprescriptive models: many of those who adopted the 'new' physical and mathematical approaches to be discussed in Chapter 8 were erstwhile Davisians by descent.

One noteworthy example of academic descent, ultimately with dissent, is Douglas Wilson Johnson (1879–1944) a key Davisian disciple, amongst whose last PhD students was Arthur N. Strahler (1918–2002), destined to lead the Counter Reformation which brought down the Davisian empire in America and Britain. A similar British case would be the story of Denys Brunsden's (1936–) and John Thornes's (1940–) reaction against the strongly Davisian ideals of their Professor at King's College, London, Sidney W. Wooldridge (1900–1963). (See Beckinsale and Chorley, 1991, p. 231 and Chapter 5.)

So it is a matter of real significance that Davis held a post at Harvard from 1878 until 1912 (as instructor in geology, 1878–85; assistant professor, 1885–90; professor in physical geography 1890–9; and Sturgis-Hooper professor in geology, 1899–1912; Chorley *et al.*, 1973, p. 852). In 34 years, in any academic institution, literally hundreds of students pass through the system, although there is some evidence (Chorley *et al.*, 1973, p. 431) that Davis was in many ways a harsh and unpopular class teacher with, in consequence, small undergraduate classes. Despite early difficulties presumably of this type, which led to a (veiled) threat that he would not be given tenure – the letter from Harvard's President Eliot on 1 June 1882 is quoted by Chorley *et al.* (1973, p. 131) – Davis recovered and, whilst teaching classes in both physiography and meteorology, began to gear up his truly phenomenal output of publications (Chorley *et al.*, 1973, appendix III). The printed scientific paper was, by the last quarter of the nineteenth century, emerging as the key source for disseminating information. Darwin, in contrast, relied largely either on books, private letters or on 'general' Journals such as *The Gardener's Chronicle*: by his death in 1881 the move towards the modern number and range of professional journals was well-established. Davis took full advantage of this proliferation of outlets. Chorley *et al.* (1973) catalogue more than 600 books, articles and reviews by Davis (in French, German and Italian as well as English) whose subject matter ranges from 'Wasp stings' (1888) to 'Airplane views of the Alps' (1921). Three of Davis's papers, including a very important and lengthy discussion of 'Sheetfloods and streamfloods' (1938) appeared posthumously.

Further, the disciples and the formidable output were linked by Davis both to extensive travels – reminiscent of Charles Lyell's magisterial forays to North America in the mid-1800s (Wilson, 1999) – which allowed The Word to be spread directly and, crucially, to a clear belief that The Word was The Truth and thus self-evidently deserved to be widely recognized as such.

However, one of the more astonishing products of Davisian old age (as noted in Chapter 1) was his 1926 paper entitled 'The value of outrageous geological hypotheses' (see Kennedy, 1983). Although he believed himself to be a devoted follower of T.C. Chamberlin's Method of Multiple Working Hypotheses (see Chapter 1) there were those who would have failed to detect its universal presence in his approach to criticism – of himself or of others. However, in the 1926 paper, he performs a quite breath-taking *volte-face* and calls for earth scientists to sit so lightly attached to their theories that their abandonment causes no mental anguish . . .

But it was, before this, in a thoroughly modern mode, reminiscent of an advertizing campaign (Chorley *et al.*, 1964, p. 634), that Davis ensured his ideas had a global and profound impact.

It helped, of course, that the ideas appeared to be – fashionably – 'Darwinian' (see Stoddart, 1966, 1986; Kennedy, 2004).

WHY 'DARWINIAN'?

When Darwin died, in 1881 his evolutionary ideas were in something of a muddle (see Ruse, 1999; Browne, 2002): the significance of Natural Selection had proved difficult to uphold in the absence of knowledge of the key mechanism of random variation provided by Mendelian genetics. What had impressed itself upon the educated mind were the concepts of inexorable and directed change over time (culminating in *Homo sapiens*), and of the survival of the fittest (a phrase coined, not by Darwin, but by Herbert Spencer). If you add these two ideas together, and you toss in the famous aphorism of the great German embryologist, Haeckel, that 'ontogeny recapitulates phylogeny' (i.e. that the developmental stages of the embryo and the growing organism trace out the key episodes in the evolution of the species: see Gould, 1977), then you can see that a theory which linked sequence to inevitable progress, in any field, was apt to be a winner. At its simplest, that was what the Davisian cycle did (see the next section). It took the formless lump of uplifted terrain – akin to the amorphous blob of the early foetus – and traced its necessary progression through a sequence 'from cradle to grave'. Just as one could expect to tell a child from an adult from a nonagenarian, so one could determine a landscape's stage in the progress from uplifted block to near-featureless plain.

Stoddart (1966; but see also Livingstone, 1992) has pointed out how widespread these pseudo-Darwinian, historical sequences became at the end of the nineteenth and beginning of the twentieth century. This was true in both the natural and the human realm. In the former, there was the concept of plant succession, from bare ground or open water to 'climax' forest. There was the development of the characteristic 'mature' zonal soil from disparate parent material, and even, one might add, the development of the full-blown mid-latitude depression from an initial difference of atmospheric temperature and pressure (see Barry, 1967).

Ironically, although Darwin's ideas were emphatically *non*-progressive and whilst he stressed the significance that chance variations imparted to the development and ramification of different species (see the only diagram provided in *The Origin*: 1859, after p. 513), the *pseudo*-Darwinian models stressed the equifinality of outcome which overrode all original differences (That is, regardless of initial circumstances, there would emerge a peneplain or a podsol, or whatever). These new models – admittedly in their simplest forms – became mechanical in their operations. This might be an imperfect way of categorizing the phenomena concerned but, like all really good didactic aids, it helped to make sense of the increasingly carefully described (and, of course, mapped and photographed) hence complex natural world. This was Davis's stroke of genius.

THE CYCLE(S)

What every student of geomorphology probably wants, at least at some stage in their development, is to be able to look at some piece of landscape – or, indeed, a single landform – and say 'I know why that looks as it does'.

When Davis began his career as a university teacher, there was no governing principle which would allow one to convey a sense of order when viewing landscapes. Dana had made attempts at classification which were of the 'hill with flat top' 'hill with pointed top' genre. Powell, as we have seen, had produced both a genetic classification of streams, and the concept of lower topographic limit to erosion, the base level. He had introduced the key idea of a balance between endogenetic and exogenetic forces. He had also described major and dramatic unconformities in the walls of the Grand Canyon – see the double page illustration, pp. 396–397, in his 1895 book. Such unconformities – as with Hutton's Jedburgh example (Figure 2.2) – must once have been widespread, nearly flat subaerial surfaces, presumably at or near base level. Gilbert had added a major idea of a balance between erosion and deposition, linked in some way to a smooth gradient, which he termed grade. But this hardly produced order from the (apparent) chaos of (say) the New England or the Appalachian landscape. Davis's genius was to impose order by resorting to what is a basic narrative template. Just as it is impossible to know why James Hutton and John Playfair saw a fluvial cycle at work in the glaciated landscape of Scotland, so it is, I think, unknowable why a man who trained as a mining engineer and who worked and lectured on meteorology, opted for an explanatory device which eliminated not only all physics and mathematics but, along the way, all true considerations of landform processes and also the major consequences of climate change.

Davis came from a staunch Quaker (and Abolitionist) background. Chorley *et al.* (1973) argue that this was, in part, responsible for his rather rigid approach to the right and wrong of academic discourse. Undoubtedly his training from the age of 16 in the new Lawrence Scientific School of Harvard, followed by a Master's course in Mining Engineering, would scarcely have added *laissez-faire* thinking to one who had, apparently, always been reserved and who devoted his leisure time to mineralogy, entomology and astronomy (Chorley *et al.*, 1973, pp. 33–34). He spent the period after graduating working in an astronomical – and meteorological – observatory in Argentina: this laid the foundations for his major contributions to meteorology culminating in his 1894 textbook *Elementary Meteorology*.

If ever there was someone who might be thought to welcome a geophysical approach to the study of landforms and to accord a key rôle to the exogenetic, climatic forces, you might expect it to be someone with Davis's background. But you would have been disappointed.

From Powell's three, precisely defined controls of the relative rate of erosion – relative relief, rock hardness and precipitation – Davis developed a comparable but far vaguer trilogy: structure, process and stage (or time). By *structure*, he included both the type of rocks (their lithology) and their folds and faults, attitude and altitude. By *process*, he simply envisaged a dominant and generalized set of the exogenetic forces: 'normal' erosion by rain and rivers was, he thought, predominant (but there were subsequently admitted to be distinct arid, glacial, limestone and coral reef process suites). By *stage*, Davis introduced the key and immensely seductive explanatory variable of relative development: youth, maturity and old age. The last stage, when relief was reduced to a gently sloping surface representing the base level of erosion, was termed a peneplain. Residual hills rising above the plain were named monadnocks, after Mount Monadnock, New Hampshire (cf. Figure 7.1).

Here is Davis's account, from his 1899 version, reprinted as chapter XIII in D.W. Johnson's edited *Geographical Essays* (1909).

'All the varied forms of the lands are dependent upon – or, as the mathematician would say – are functions of three variable quantities, which may be called structure, process and time. In the beginning, when the forces of deformation and uplift determine the structure and attitude of a region, the form of its surface is in sympathy with its internal arrangement, and its height depends on the amount of uplift that it has suffered. If its rocks were unchangeable under the attack of external processes, its surface would remain unaltered until the forces of deformation and uplift acted again; and in this case structure would be alone in control of form. But no rocks are unchangeable, even the most resistant yield under the attack of the atmosphere, and their waste creeps and washes

Figure 7.1 *What a Davisian peneplain would ideally look like: looking over the Potomac River towards Sugar Loaf Mountain, Maryland. (From Shelton, 1966, figure 146.)*

downhill as long as any hills remain; hence all forms, however high and however resistant, must be laid low, and thus destructive process gains rank equal to that of structure in determining the shape of a land mass. Process cannot, however, complete its work instantly, and the amount of change from initial form is therefore a function of time. Time thus completes the trio of geographical controls, and is, of the three, the one of the most frequent application and of a most practical value in geographical description' (1899: in Johnson, 1909, p. 249).

The sequence of events was laid out even more clearly in the 1904 article entitled 'Complications of the geographical cycle' (in Johnson, 1909, chapter XIV).

'In the scheme of the ideal geographical cycle a complete sequence of landforms of one kind or another may be traced out. The cycle begins with crustal movements that place a given land mass in a certain attitude with respect to base-level. The surface forms thus produced are called initial. Destructive processes set to work upon the initial forms, carving a whole series of sequential forms, and finally reducing the surface to its ultimate form, – a low plain of imperceptible relief. The sequential forms thus constitute a normal series by which the initial and the ultimate forms are connected. As a result, the sequential forms existing at any one moment are so largely dependent on the amount of work that has been done upon them that they are susceptible of systematic description in terms of the stage of the cycle which they have reached. Moreover, the correlation of all the separate forms appropriate to any one stage of the cycle is so intimate and systematic that any single form may be designated in an appropriate and consistent terminology as a member of the group of related forms to which it belongs, and thus, better than in any other way, the features of the lands may be systematically and effectively described' (1904: in Johnson, 1909, pp. 279–80).

This version was being read to the Eighth International Geographical Congress, in Washington, where Davis evidently suspected that there might be the unconverted or even sceptics in the audience and he deals with some key objections in this way:

'The suggestion has been sometimes made that a scheme having a less proportion of imagined or deduced elements and a greater number of actual examples would be more generally acceptable to geographers. In reply it may be said that the scheme of the cycle is not meant to include any actual examples at all, because it is by intention a scheme of the imagination and not a matter of observation; yet it should be accompanied, tested, and corrected by a collection of actual examples that match just as many of its elements as possible' (1904: in Johnson, 1909 p. 281).

To this basic cycle, Davis added the 'Complications' of the paper which included: slow rather than rapid uplift; interruptions due to tectonic movements; special agencies namely ice and wind; and 'climate changes independent of the ideal cycle . . . noted by the semi-technical term "accidents"' (1904: in Johnson, 1909).

'It thus appears that the scheme of the ideal cycle may be gradually and systematically modified until its deductions cover all manner of structures, agencies, waste forms, interruptions, and accidents. When thus conceived it is a powerful instrument of research, an invaluable equipment for the explorer. It is not arbitrary or rigid, but elastic and adaptable. It is a compendium of all the pertinent results of previous investigations' (1904: in Johnson, 1909, p. 292).

Or, you may say, how to have your cake and eat it . . .

I trust these excerpts show why the Davisian ideal – which is not really a cycle, but a sequence – proved both so attractive and so infuriating. In common with all great deductive and didactic schemas, *if* you accept the principles, then the rest follows. This, in turn, makes it well-nigh impossible to *test* the propositions (although we shall see that A.N. Strahler tried: Chapter 8). Since land is, indubitably, uplifted and since there are extensive erosional unconformities which must have represented surface forms something very like peneplains and since Davis allows for all manner of 'accidents', then it seems self-evident that The Cycle can provide the key to the history of any landscape.

Of course, as Davis himself and his close students showed, it is actually very difficult to unravel what came to be known as the denudation chronology of any portion of the real world. Davis's 1889 study of 'The rivers and valleys of Pennsylvania' (in Johnson, 1909, chapter XIX) gives a taste of the problems. For an exhaustively complete discussion, see Chorley *et al.* (1973, chapter 11, *et seq.*).

It is also significant that, by the early twentieth century, Davis is addressing *geographers*, rather than geologists. This distinction was to become variously important in different parts of the world, as the territories of academic disciplines became more firmly delineated. The term 'geomorphology' seems (Chorley *et al.*, 1964, p. 615) to have entered general usage via the American geologist W.J. McGee (1853–1912) in the 1880s. As Chorley *et al.* state:

> 'Before the early nineteenth century [ideas relating to the development of the physical landscape] formed the bulk of geology, but thereafter, largely because of the rapid growth of stratigraphy and palaeontology . . . occupied a decreasing share of that science, although [their] fundamental necessity in general geological knowledge was never in doubt. After about 1860 the study became part of both geology and physical geography and was later also known as physiography or geomorphology. Today it forms the link between geology and geography and probably its association with the latter discipline is one of the main reasons why it has failed to develop along more strictly scientific lines' (1964, p. xi).

By 'scientific' Chorley, in particular, is thinking in terms of the reductionist approaches which developed post World War II (see Chapter 8). And there can be little doubt that Davis exerted all his powers to create a separation between the aspects of landscape development which should be the provenance of physical geography and those of geology. In 1900, Davis wrote: 'It may be urged that in many geological discussions from which geography has taken profit, consideration is given to form-producing processes rather than to the forms produced. This was natural enough while the subject was in the hands of geologists; but geographers should take heed that they do not preserve the geological habit. The past history of landforms and the action upon them of various processes by which existing forms have been developed, are pertinent to geography only in so far as they aid the observation and description of the forms of today' (in Johnson, 1909, p. 83). (For recent, contrasting discussions of the differing approaches of geography, geology and geophysics, see Church, 2005; Summerfield, 2005: see also Giusti, 2004.)

The relegation of process to a subsidiary (and superficial) rôle in explanation may well account for Davis's apparently extraordinary resistance to incorporating what was becoming a well-documented case for repeated, wholesale, climatic shifts producing major continental ice advances and retreats during the most recent geological period, the Pleistocene. The establishment, in 1909, by the Germans Albrecht Penck (1858–1945) and Eduard Brückner (1862–1927) that there had been at least four great ice advances in the Alps (Beckinsale and Chorley, 1991, p. 45) was evidently known to Davis, who was – for many years – on warm personal terms with Penck. Associated with such glacial advances there must have been both eustatic and isostatic changes in sea level, i.e. in base level. How far could one expect the 'normal' – or any other – cycle to run a simple course in such an unstable world? It is possible that – in the continuing absence of absolute dating – Davis considered that Pleistocene events were mere transient episodes in the geological story. He was certainly not the last geomorphologist to assume that processes marched on to a characteristic conclusion despite all the 'accidents' of the Quaternary.

It is impossible, here, to survey the sweep of Davis's contributions. What was central was his invention of a lucid and apparently comprehensive outline of the sequential progress of landforms, given specific geological settings and specific processes at work. He produced this vision despite a strong background in mathematics and physics, despite a close working relationship with G.K. Gilbert and despite growing evidence that the world's recent history was much less stable than his *schema* really required.

THE REACTION

No major scientist is ever likely to produce a single, seamless vision of his or her subject matter. Davis, certainly, changed his mind about both tectonics and arid landscapes, late in his life (although this made little impact on the general reception of his ideas). But it is crucial to realize that he was by no means the only Grand Old Geomorphologist around, even by 1900, and, although there were those who accepted the value of The Cycle, in whole or in part, there were equally those who considered it irrelevant or decried it as pernicious. This variable response was true even within North America, but far more marked in western Europe: especially in Britain, France and Germany (see Beckinsale and Chorley, 1991).

First, there were those who adopted The Cycle and its sequential forms as the key to unravelling landscape history. These Denudation Chronologists – or peneplain spotters as they were later unkindly termed – were, of course, terribly handicapped by the absence of any method of absolute dating of either deposits or surfaces until the 1950s. Nevertheless D.W. Johnson in the Appalachians, S.W. Wooldridge in southeastern England (cf. 1936), Henri Baulig (1877–1962) in the Massif Central of France (see Masutti, 2002) and Charles Cotton (1885–1970) in New Zealand (cf. Cotton, 1945) extended and developed the Davisian approach to a perhaps surprising range of landscapes.

Then there were those who may perhaps have taken their cue from Davis's development of 'special cases', notably those linked to different climatic régimes. The turn of the twentieth century had seen a massive expansion of European scientific travel in Africa, in particular. Out of this in part emerged a strong, largely European group of Climatic Geomorphologists, amongst them the Germans Siegfried Passarge (1866–1958), Herbert Louis (1900–1985) and above all Julius Büdel (1903–1983); and the French Pierre Birot (1909–1984), André Cailleux (1907–1986) and Jean Tricart (1920–2003).

Third, there were those who considered the Davisian model so fundamentally flawed that they wished to replace it with a different, albeit similarly deductive and didactic model of their own. First of these was the young Walther Penck (1888–1923) (the son of Albrecht) who in 1924 published a lengthy paper that was finally translated into English only in 1953 (by H. Czech and K.C. Boswell) as *Morphological Analysis of Landforms*. In sharp contrast to Davis's basically simple (or even simplistic) views on the rôle of tectonics as the mere precursors of landscape development, Walther Penck – who had worked extensively in the Andes and been as struck by Darwin with the intensity of their endogenetic forces – produced 'a tectonically-controlled philosophy' (Beckinsale and Chorley, 1991, p. 121 and chapter 10). The younger Penck had also spent time in North America and been introduced to G.K. Gilbert, amongst others: there is certainly a distinctly Gilbertian air to some of his assumptions about the significance of slope gradient, especially with respect to debris production and removal (Beckinsale and Chorley, 1991, p. 357). It is, however, probably fair to say that – partly as a result of an 'explanation' of Penck's scheme by Davis which was fundamentally mistaken – the importance of this tectonically dominated alternative to Davis's historical narrative had rather limited impact, especially in the English-speaking world.

The other major challenge to the Davisian Cycle came in 1951, when Lester King (1907–1989), published *South African Scenery*. King had been one of Charles Cotton's students in New Zealand and then moved to the University of Natal. Given the Davisian background, it is, as Beckinsale and Chorley note (1991, pp. 190–1) surprising to hear King's total rejection of the Master's views in his later *Morphology of the Earth*, thus:

> 'The classic account of the "Normal Cycle of Erosion" as expounded by W.M. Davis has proved regrettably in error. With its emphasis on universal downwearing, it was a negative and obliterating conception resulting from cerebral analysis rather than from observation, and has led to sterility in geomorphic thought and retarded progress in the subject severely . . . We accept Davis's original idea of a cycle of erosional changes in landscape while rejecting the method of landscape development which he advocated' (King, 1962, pp. 162–3, quoted by Beckinsale and Chorley, 1991, p. 190).

Just as Walther Penck's model reflected his experience in the tectonically active Andes, so King took as his basis the great plains and escarpments of southern and eastern Africa. The apparent dominance of cliff retreat – backwearing – would, of course, have seemed entirely sensible to Powell or any of those whose work in the USA canyonlands had demonstrated the reality and extent of scarp retreat (recall Figure 6.3). Perhaps there really *are* different underlying modes of landscape evolution in different climatic settings . . .

And, finally, there was a not insubstantial international group who simply thought the whole Davisian structure overblown and totally futile. This included some USA workers – including R.S. Tarr (1864–1912), a Davisian student who moved to Cornell and had a fairly comprehensive falling out with The Master, and the Chicago School of T.C. Chamberlin and R.D. Salisbury (1858–1922) – but perhaps the most devastating criticisms came from the German, Alfred Hettner (1859–1941) of Heidelberg University.

'Davis's cycle theory . . . is not in fact the general theory of landscape development for which it sometimes passes, but it is limited to the influence of uplift . . . Factually, this theory is not as original as has often been thought. It introduces only a novel form of expression . . . Distinguishing cycles has become a favourite aim of geomorphological studies. Indeed, one can almost say that the setting up of cycles has become a mania . . .

Davis's school has whole segments of the earth's surface moving up and down, being destroyed and levelled, as though they were stage scenery . . . [H]ere again, the theory is usually not supported by the facts. Most "peneplains" are postulated on slender evidence . . .

To characterize landforms by their age leads to error . . .

While Davis talks a lot about "life", his scheme lacks vitality, the landscape picture it gives has a moribund and dismal emptiness . . . The unending variety of rock types and the way they are arranged is submerged beneath the schematic contrast of "hard" and "soft" . . . [H]e considers the diversity of climatic influences in much too cursory a fashion . . .

I can see Davis's approach only as an episode, not as a step forward in geomorphology. Its simplicity and the energy of its advocates has rapidly won for it a wide circle of adherents; it has enlivened geomorphological research and led to a number of correct results. But as a whole it has been abortive, and studies founded upon it have produced many failures. A lot of debris has to be cleared away to reopen the field to unrestricted research' (Hettner, 1928, quoted by Beckinsale and Chorley, pp. 510–2: see also Tilley, 1972).

But, the Davisians would reply, What was the alternative?

Nevin Fenneman (1865–1945) somewhat inadvertently gave the answer in a passage which several generations of R.J. Chorley's Cambridge students could chant as proof that the Quantitative Revolution had displaced Davis's ideas:

'Cycles have parts and the parts make wholes, and the wholes may be counted like apples. Non-cyclic erosion can only be measured like cider. There is neither part nor whole, only much or little.' Fenneman (1936) quoted by Beckinsale and Chorley (1991, p. 196).

Indeed . . .

Chapter 8

Reinventing a Newtonian Universe: the Reductionist Revolution, 1945–1977

'Playfair did pioneer work based on ocular observations. There were available to him neither the results of measurements nor the hydrophysical laws necessary to their quantitative interpretation. It appears that the time has now come when such a quantitative interpretation can be undertaken' (Robert E. Horton, 1945, p. 280).

INTRODUCTION

I think there is little doubt that the 1945 paper by the American hydraulic engineer, Robert E. Horton (1875–1945) is the crucial springboard for what I would term the 'Reductionist Revolution' which arose in Anglo-American geomorphology in the mid-twentieth century. This influenced the work on fluvial and slope processes in particular (see below). However, it is necessary both to recall that the Newtonian emphasis on measurement and physical experiments and laws goes back to Buffon and his red-hot metal spheres, to Hutton with his experiments undertaken with Sir James Hall on heat and pressure (see Repchek, 2003) and, of course, to Darwin whose earthworm investigation remains one of the longest and most detailed quantitative geomorphological studies.

However, what was played out in the earth science community after 1945 was a continuation of a long-running conflict. On the one hand, those who taking an approach descended from strict Newtonian mechanics ('Action and reaction are equal and opposite') and the Leibnitz/Newton invention of calculus, opted for a reductionist, experimental approach to landforms and, especially, landforming *processes*. On the other, those who – seeing the historical and both temporally and spatially contingent nature of the connection between form and process (see Simpson, 1963) – argued that a reductionist view of process was a necessary but not sufficient condition to explain earth surface features, or, indeed, *anything* with an important history. As the great German–American biologist Ernst Mayr has put it:

'When part of a historical narrative consists of functional processes, they can be tested by experiment. But the historical sequence as such . . . can be reconstructed only on the basis of inferences derived from observations . . . It would

be interesting to go through the history of science and see how often a mis-
placed insistence on experiment has caused research to move into unsuitable
directions' (1982, p. 856).

But before we look at the geomorphological reaction to W.M. Davis's emphasis
on stage, which led to an antithetical but a similarly narrow focus on process,
there are four major developments in earth science generally which require
notice: the development of (we think) accurate absolute dating methods; the
acceptance of Continental Drift and the invention of plate tectonics; the recognition of the frequency and intensity of climatic shifts, especially in the Quaternary;
and the ability to examine the forms and processes of our nearest neighbours in
the solar system.

ABSOLUTE DATING

When Rutherford suggested that the source of heat represented by radioactivity
(see Chapter 2) would allow the Earth to be seen as much older than a simple,
cooling body, he also opened the door to the development of what is now a
whole host of absolute dating techniques. The most basic of these use the principle of radioactive decay as a more or less accurate clock. Each radioactive element
loses one half of its radioactivity, by an apparently random process, in a fixed
period (known as its half-life). The half-life may be extremely long (as with the
Uranium series) or extremely brief (Caesium): see Roberts (1998, p. 27). There is,
then – providing suitable datable materials exist – now a range of ways in which
earth substances can be given an age in (more or less) calendar years. It is this
basic technique which Patterson, in 1953, used to calculate that the age of the
Earth was just under 4.6 thousand million years (see Holmes, 1965, pp. 378–380;
Repchek, 2003, pp. 202–4; Bryson, 2003, pp. 138–142).

Although all dating techniques have pitfalls, it is really impossible to underestimate what a fundamental change they have made to earth science. Rutherford
and his colleague Soddy really *did* create a basic paradigm shift. Without this, the
present book would never have been written, since – if we were still accepting a
simple Newtonian account such as Kelvin's – we would be trying to relate our
observations of process and form (as well as organic evolution) to a reduced
timescale of, say, 20 million years. It is extremely salutary to pause and consider
what immense shifts in our thinking would be needed were that to be the dominant view.

Once we can date either deposits or, increasingly, surfaces (see Chapter 9),
not only can we hope to construct a real denudation chronology, but we can
also try and obtain an estimate for the rates of operation of processes which is
far more accurate than Darwin's guestimate about the Wealden Denudation.
Most crucially, we can create a chronology which incorporates the crustal processes of mountain building and plate movement as well as the shifts of climatic
belts.

PLATE TECTONICS

The apparent mirror-image of the continents on either side of the Atlantic ocean has attracted attention since the seventeenth century. It was, however, the German meteorologist, Alfred Wegener (1880–1930) who was the originator of the modern vision of what he termed 'continental displacement' that became 'continental drift' in English (see Wegener, 1924; Beckinsale and Chorley, 1991, pp. 19–30).

As T.S. Kuhn (1962, p. 146) pointed out, 'All historically-significant theories have agreed with the facts, but only more or less' and this was most certainly true of Wegener's ideas. Most crucially, the evidence he proposed for the rate of continental displacement was simply and wildly inaccurate (he was suggesting metres a year: by 2004, scientists were thinking more in terms of a maximum of 20 cm a year: Symons, 2004, p. 793). He also freely admitted that he could not explain *how* the continents moved: 'The Newton of drift theory has not yet appeared' (Wegener, 1929, in Biram, 1960, p. 167). But Wegener was absolutely clear that the observational evidence of comparable geological structures, as well as plants and animals on either side of the Atlantic could only be explained by erstwhile continuity, i.e. before the opening of the Atlantic. It was similar observations, both of the magnetic striping on either side of the Mid-Atlantic Ridge and of the increasing age of the volcanic islands as one travels away from the Ridge, which finally led to the geophysical community accepting continental motion, from the early 1960s onwards.

The story of the earlier rejection of the whole notion of continental drift by the North American geological community has been brilliantly recounted by Naomi Oreskes (1999). She provides fascinating evidence (as noted above, Chapter 1) that it was that community's rigid adherence to T.C. Chamberlin's Method of Multiple Working Hypotheses which proved the crucial stumbling block: there was no scope to accept a new paradigm, it had to be treated like a singular and therefore reprehensible Ruling Hypothesis. Moreover, the whole of Wegener's idea smacked of the eighteenth century Catastrophist notions and seemed to fly in the face of the sound Uniformitarian principles firmly established since Hutton, Playfair and Lyell (cf. Oreskes, 1999, p. 200). Ruse (1999) has made an elegant consideration of the competing methodologies of Kuhn and Karl Popper (1959). The latter, with its emphasis on the necessity that hypotheses be testable and thus potentially refutable is very close to Chamberlin. As noted in Chapter 1, Ruse concludes that one can only apply the Chamberlin/Popper method to cases of the 'puzzle solving' undertaken during Kuhn's periods of 'normal science', but *not* to cases of revolutionary new theories.

However, Wegener's vision was, ultimately – and very rapidly – accepted in its modern form in the 1960s, with the result that one now had to contemplate not merely changing continental elevation (cf. Lyell) as relevant to climatic shifts, but also the prospect that continents could migrate in and out of different climatic zones. Not, however, that those were now seen as fixed.

CLIMATE INSTABILITY

The Earth is a slightly squashed spheroid, with an axis of rotation which is currently tilted at 23 1/3° from the vertical. As a consequence, the overhead sun appears to swing between the Tropic of Cancer (23 1/3°N) and the Tropic of Capricorn (23 1/3°S) from one solstice to the next. Sunlight is experienced for a full 24 hours at the Arctic and Antarctic Circles (66 2/3°N and S) at the summer solstice and there is complete darkness at the winter solstice. Between the polar and tropical latitudes there are more or less parallel climatic belts: cool temperate, warm temperate, the west coast Mediterranean case of the semi-arid zone, the hot deserts, the seasonally wet (savannah) tropics and the humid equatorial regions. As we have seen, by 1945 many geomorphologists had been endeavouring to link landforms to climatic zones, developing the notion of morphoclimatic regions (see Beckinsale and Chorley, 1991, pp. 436–455).

We have discussed, in Chapter 4, Lyell's objections to James Croll's ideas of cyclic shifts in the three variable components of solar insolation reaching the Earth's surface. Croll's ideas were later elaborated and refined by the Serbian astrophysicist Milutin Milkankovitch (1879–1958: see Roberts, 1998) and have been widely accepted as identifying the 'pacemaker of the Ice Ages' (Imbrie and Imbrie, 1979).

There are, in fact, thought to be two separate mechanisms operating in producing global climatic shifts (see Bowen, 2004).

First, from time to time, the configuration of the Earth's land masses apparently creates circumstances conducive to the development of polar ice caps. Once these have formed, then the Croll–Milankovitch mechanisms appear to create a regular sequence of cold and warm episodes. We are, we think, at present in a warm interglacial and may therefore expect a return to cold, glacial conditions within the next 1000 years or so. Probably . . . In the 1960s, this was widely predicted as imminent. In the 2000s, there are those who consider that human-induced, enhanced global warming has swamped the insolation pattern and led to the end of the Quaternary Ice Age. We have no way of knowing which is the case although we may be able to gain some useful evidence from Mars (see below).

What is indisputable is the fact that the global climate zones have been expanding and contracting in extremely complex ways for most of the past 2–3 million years. There are variable estimates for just how many fluctuations, but certainly more than 50 have been recognized (Bowen, 2004). This picture of wholesale climatic – as well as tectonic – instability poses a very important question: how long does it take for 'characteristic' landforms to develop? If at all?

EXTRA-TERRESTRIAL EVIDENCE

Since the first men walked on the Moon, in July 1969, the Earth has ceased to be a sample of one. With the manned and unmanned probes to the Moon,

Venus, Mercury, Mars and the moons of Jupiter and Saturn, we can begin to establish both similarities and contrasts between our planet and its neighbours (see Baker, 1993).

It is clear that, in comparison with Mercury and the Moon, our planet's surface is renewed so frequently that there are virtually no traces of meteorite impacts. However, it is not entirely evident why the processes of plate tectonics – which we think are responsible for this recycling of surface materials on Earth – may not have operated on Venus and Jupiter's moon Io; but Mars also seems to possess distinctive fluvial landforms (and its canyons, because of the lower gravity, make the Grand Canyon look puny): what made them? (see Bandfield *et al.*, 2003; Malin and Edgett, 2003). The recent (January 2005) evidence of 'fluvial' forms on Saturn's moon Titan, cut by liquid whose origin is uncertain (Kerr, 2005) adds to the geomorphologist's range of examples requiring explanation.

Finally, there is evidence that the Martian ice caps show banding, suggesting that there have been different episodes of ice growth and retreat. These must be linked to variations in the Martian receipt of insolation. If we can sample and date the layers, then we can create an independent Martian Milakovitch chronology which would provide a firm test of the reliability of our present theory as regards Earth.

One way or another, the Earth as we see it in 2005 is a much more complicated entity than was remotely envisaged even in 1945.

R.E. HORTON AND HIS PRECURSORS

When Robert Horton's (1875–1945) posthumous 95 page scientific paper appeared in 1945, it was aiming to do a number of things. First, as the quotation at the head of this chapter emphasizes, it aimed to provide a quantitative justification for Playfair's qualitative vision of the role of (fluvial) processes. Secondly – and perhaps rather startlingly to those brought up to see Horton as the Davisian antithesis or nemesis – it was to provide an explicit, quantitative basis for the Davisian 'Normal Cycle' (see Kennedy, 1992). Third, it was trying to create a comprehensive reductionist model that would describe and account for the development of the fluvial *landscape*.

Now, there had been a long tradition of physical and mathematical investigations of both processes and, to a lesser extent, forms. These are documented by Chorley *et al.* (1964), where each of the four chronological sections concludes with a discussion of 'quantitative' studies.

The first of these were concerned with hydraulics and river mechanics (Chorley *et al.*, 1964, p. 87) and although one must accept that Guglielmini (1655–1710) was the first European to produce a scientific analysis of river flow, in 1697, one does need to recall that the great ancient hydraulic civilizations – notably that of China (see Needham, 1971, section 28) must have had a similar or greater grasp of fluvial mechanics. The foundations of hydraulics were generally established in the eighteenth century (see Rouse and Ince, 1957) by largely French scientists: Pitot, Daniel

Bernoulli, Bossut, Chézy and Du Buat. Buffon (1749, pp. 340–351) has a lengthy discussion of basic processes. One may consider that much of the work of L.B. Leopold and his co-workers (see below) consisted of reintroducing these well-established principles to the non-engineers concerned with rivers.

Rivers continued to attract quantitative attention throughout the nineteenth century. The French civil engineer A. Surell was of particular significance, producing a major work on Alpine streams in 1841. A decade later (1851) saw the appearance of an important volume by the English engineer, T.J. Taylor, which is often distinctly reminiscent of American fluvial geomorphology 100 years down the line. One must also mention the 1861 report by the American Army engineers, Humphreys and Abbott, on the Mississippi River; and the contribution of a basic velocity equation by the Irish engineer, Robert Manning, in 1890. Rouse and Ince (1957, p. 187) note that the dimensionless number now attributed to the Englishman William Froude was one 'which he never even used.' Perhaps the culmination of the quantitative study of rivers was G.K. Gilbert's enormous paper (with assistance from F. Matthes) on 'The transportation of debris by running water' (1914).

The purely fluvial is, in many senses, a very simple geophysical system – provided you are cautious about the boundary resistance of the channel (recall Figure 6.5, Chapter 6).

A similarly 'simple' problem is ostensibly that of the retreat or evolution of bare cliffs and there had been, by 1945, two major mathematical treatments, by the English clergyman, Oswald Fisher, in 1866, and by the Western geologist, Clarence Dutton, 1882 (see Chorley et al; 1964, especially figure 123, p. 585). Now one can devise all kinds of neat mathematical models which allow one to manipulate the profile form of a slope: it is really comparable to fixing two ends of a piece of string and seeing what variable shapes can result. It becomes more complicated if you allow the location of the slope base or top to vary over time and the whole nineteenth century exercise has an enthusiastic and exhaustive descendent in the Austrian, Adrian Scheidegger's *Theoretical Geomorphology* of 1961 (later editions, 1970 and 1991).

Evidently the non-Davisian landscape models of Walther Penck and L.C. King focused in large measure on sequences of slope evolution. However, before Horton's 1945 paper one might argue that the only wholesale attempt to create a quantitative landscape model was made by the French authors General Gaston de la Noë and Emmanuel de Margerie in 1888 (see Chorley et al, 1964, pp. 627–34). *Les formes du terrain* was in many ways a precursor of Horton. It was not merely a thorough-going Fluvialist explanation of landforms, but a fundamentally quantitative one. In their first chapter where (p. 5) the fluvial pattern is described as '*le modelé normal*', they put forward three key observations which convince them (à la Playfair) of the dominance of fluvial processes: the continuity between slopes and streams; the regular increase in size of river channels and their ramifying form; and the general accordance of junctions in the fluvial networks. They later discuss the evolution of the profiles of equilibrium (à la Gilbert) and consider that, once this has been achieved '*le profil longitudinal n'éprouve presque plus de changement*' (1888, p. 75).

However, the impact of this work was undoubtedly limited. Horton's 1945 paper must be taken as the springboard from which the fluvial Reductionist Revolution was launched.

Horton had been fascinated by the process by which rainfall (or, more generally, precipitation) turned into runoff. He published a major study on this in 1933 in which he developed the idea that it was the ease with which infiltration occurred which governed both the rate of runoff and the location of rills and fingertip channels on slopes. This process, now termed Hortonian infiltration-excess runoff, assumes the central rôle in the 1945 paper (entitled 'Erosional development of streams and their drainage basins; hydrophysical approach to quantitative morphology'). In essence, a hillside's surface is viewed as a series of adjacent 'buckets'. When liquid water is available at the surface, it will endeavour to sink into the ground (unless the surface is impermeable, temporarily or permanently) until all the bucket's capacity is full; the excess will then run off, downslope. Evidently, the further you get from the slope head or divide, the greater the volume of water running downslope and, hence, the greater the likelihood that the flow will become sufficiently turbulent and/or impart sufficient stress to erode a rill or fingertip channel onto the slope face. Above the rill head there will be a smooth, unrilled zone, which Horton termed the 'zone of no (sheet) erosion'. This corresponds to Gilbert's zone of dominant soil creep (see Chapter 6, Figure 6.4C).

What Horton then does is translate this microprocess into the origin of fully developed stream channels and, finally, a fluvial network. (The details of his discussion of micropiracy between rills have never been entirely convincing; and we should note that where there is a soil and vegetation cover on a slope, Hortonian runoff generation is generally replaced by a more complex process where excess water is routed under the surface, as throughflow and emerges in a more sporadic pattern, often at the slope base: see Kirkby and Chorley, 1967.) It is what Horton does with this network which proved truly revolutionary. That streams may be viewed as a hierarchy is an old and obvious idea: Playfair described it very clearly:

'Every river appears to consist of a main trunk, fed from a variety of branches, each running in a valley proportioned to its size, and all of them together forming a system of valleys' (1802, p. 102).

Horton simply devised a way of numbering streams according to their relative size and location: curiously Buffon had invented virtually the same system in 1749 (pp. 353–4). Figure 8.1 shows the essence of the Hortonian stream ordering system (A) together with its simplification and modification by A.N. Strahler (B, 1952) and a mathematically more tractable version produced by R.L. Shreve (C, 1966).

In Figure 8.1A, Horton has constructed a hierarchy on the same basis as that shown in 8.1B, but has then relabelled whole streams by the order at their mouths. So the Third Order basin has a single 3rd order stream, a single 2nd order stream and five 1st order ones. It turned out, in practice, to be extremely difficult to decide which channels were the main ones, hence Strahler's version, which shows one, short 3rd order stretch, two 2nd order and seven 1st order. It is evident

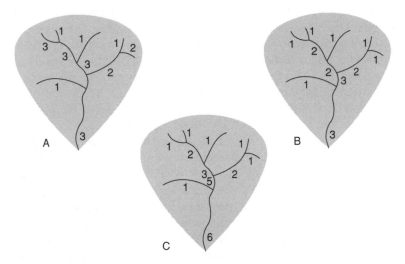

Figure 8.1 *Stream ordering. For explanation, see text.*

(Figure 8.1B) that one of the two 2nd order channels has four tributaries and the other only three, i.e. they are not exactly equivalent, although they have the same rank. This creates mathematical problems, resolved by Shreve's system (Figure 8.1C) where each tributary is counted to produce the basin magnitude. The advantages of the Horton/Strahler scheme are that it is relatively easy to calculate any basin's order and it is also the case that 4th order networks are particularly common (the Mississippi is only probably 10th or 12th order (Leopold *et al.*, 1964, p. 141), so conservative is the hierarchical structure). This proved valuable as a starting point for comparative studies. (Finding networks of equivalent magnitude is more difficult.)

Horton went on to analyse the geometrical properties of his ordered networks and established what he considered to be a number of Laws. First, that the number of streams of different orders approximated to a geometric sequence, with a ratio of 2.5 between them. Second, that there was a geometric increase in average channel length from 1 upwards. Third, there was a geometric decline in channel slope with order. There are two possible snags with this. The first results from the fact that Horton drew his relationships as lines on semi-logarithmic paper, with order on the axis. Since order is an ordinal variable and hence, as we have seen, two streams of similar order may be of different magnitudes it is possible to calculate ratios for number, but it may be thought dubious to plot graphs. Second, if you use Strahler's very different ordering, you also produce 'Laws'. This is either baffling or exciting, depending on your point of view (cf. Woldenberg, 1968; Rodriguez-Iturbe and Rinaldo, 1997).

However, what Horton did then was, ostensibly, very simple. He measured the total length of stream channel in a basin, divided it by the basin area and created an index of Drainage Density. One could now compare the degree of dissection of landscapes, it seemed, precisely. If one took three 4th order basins and calculated that Drainage Dernsity was 1, 5 and 60 km⁻¹ respectively, then it was immediately evident that fluvial erosion was dramatically more effective in the last than in the first. Although there are some real practical problems with Horton's precise definitions (see Kennedy, 1978) and whilst Drainage Density can be differently defined either for channels only, or for 'contour crenulations' (i.e. all the 'dry valleys' in the network) with interesting implications (see Gregory, 1966; Blyth and Rodda, 1973; Schumm, 1997), no one has really found a better index.

Overall, then, Horton's paper not merely seemed to incorporate Playfair's 'ocular observations' but also permitted Davisian stages to be expressed in apparently precise quantitative terms (but see Kennedy, 1992). It was, indeed, a turning-point.

Its appearance coincided with the end of World War II. That had enormously increased the investment in global mapping – especially using aerial photographs – and had, in the development of the atomic bomb, demonstrated, literally, the tremendous power of applied physics. There were two American workers who seemed to adopt Horton's approach, although they took it in somewhat different directions: Arthur N. Strahler (1918–2002) and Luna B. Leopold (1915–).

STRAHLER AND HIS SCHOOL

As mentioned in Chapter 7, Arthur Strahler was one of the last PhD students of W.M. Davis's key pupil, D.W. Johnson. After a PhD at Columbia University, Strahler ultimately became Professor of both Geology and Physical Geography there, until his early retirement in the late 1960s.

Following graduate work in the Kaibab and Coconino plateaux of the American West, Strahler decided to put the Davisian dictum concerning the relative significance of structure, process and stage to the test. Using the field area of the Verdugo Hills, California and a revolutionary combination of experimental design, field measurements of slopes and inferential statistical analysis, Strahler's 1950 paper on 'Equilibrium theory of erosional slopes approached by frequency distribution analysis' was a landmark. It demonstrated that the key to slope development was an apparently present-day, ubiquitous process – in this case the intensity of basal undercutting by a stream – and that 'characteristic' or 'equilibrium' slope angles developed in different process domains. (This work was expanded in a more carefully designed experiment by Strahler's student M.A. Melton in 1960.)

However, far and away Strahler's most dramatic impact came in 1952, when he produced a paper entitled 'Dynamic basis of geomorphology.' In this, although allowing – in passing – a rôle for historical studies, he set out in great detail his vision of a subject which would rid itself of qualitative and impressionistic views of poorly defined processes and forms and move towards the development of 'mathematical expressions to serve as quantitative natural laws'. It is, I think, difficult to underestimate the importance of this work. It encapsulates the essence of the reductionist approach which takes classical Newtonian mechanics and later thermodynamic concepts of equilibrium as its *vade mecum*.

Strahler's other major contribution to twentieth century geomorphology was made through his graduate students (see his account of the Columbia School, 1992), at least four of whom have had significant rôles to play on the global stage.

Marie Morisawa (1918–1993), a fluvial specialist (the first woman, you may note, to be mentioned in this account apart from Buffon's milkmaids) became Professor at the State University of New York in Binghamton and was instrumental (with Donald Coates) in setting up a major annual symposium – 'The Binghamton' – which has been running since 1970 and acts as a major focus for international geomorphologists.

The journal *Geomorphology* (1987–) has been a by-product of this undertaking.

Mark A Melton (1930–), a brilliant linguist and geologist from Oklahoma and Yale was undoubtedly the cleverest of the Columbia graduates. His PhD thesis (1957) on 'An analysis of the relations among elements of climate, surface properties, and geomorphology' remains almost the only wide-ranging, properly designed and analysed quantitative study on the nature of the links between process and form in fluvial landscapes. This work resulted in an extraordinary article, in 1958, which related the development of drainage basins to their evolution in an 'E_4 phase space' (see Keylock, 2003). It was largely incomprehensible when it appeared and, like much of Melton's (unfortunately rather limited) published work, was only taken up and rediscovered 30 years or so down the line. After positions at Chicago, Tucson and British Columbia, Melton – like Strahler – took a very early retirement in the late 1960s. His sole principal student, Michael A. Church (1942–), currently holds a Chair at the University of British Columbia.

Stanley A Schumm (1927–) is, in many ways, the most successful geomorphologist from the Columbia School. He began by working on the romantically named Perth Amboy Badlands (actually a landfill site in New Jersey, conveniently bulldozed after his field work) which led to a major paper in 1956 'On the rôle of creep and rainwash on (*sic*) the retreat of badland slopes'. This showed that it was the infiltration characteristics of the ground surface, determining whether surface or subsurface processes dominated, which controlled slope development: parallel retreat dominating with rainwash, slope decline with creep. Schumm worked for many years at the U. S. Geological Survey, in Denver, before moving to Colorado State University at Fort Collins. The best flavour of his work is given by his 1977 textbook, *The Fluvial System* which shows his imaginative use of observation and experiment, often with counter-intuitive results. It is well worth noting that Schumm, despite his general acceptance by late twentieth century workers as a full-blooded Reductionist and process geomorphologist has retained a strong enthusiasm for the historical approach of W.M. Davis (see King and Schumm, 1980). As with Gilbert, Schumm demonstrates the need to combine the historical with the experimental approach to problem-solving in geomorphology. One of his more interesting students, a pupil of R.J. Chorley from Cambridge, is M. Paul Mosley who has been based in New Zealand since the 1970s. If there is a heir to G.K. Gilbert, I would judge it to be Stan Schumm.

Finally, Strahler attracted an Oxford graduate Richard J. Chorley (1927–2002: see Stoddart, 1997). After a period at Brown University, Chorley moved to Cambridge in 1958, where he remained until his death.

One of Chorley's key papers was the 1962 'Geomorphology and General Systems Theory'. This took up both Strahler's 1952 call to arms and Strahler's interest in the ideas of the biologist, Ludwig von Bertalanffy (1950), who saw the thermodynamic concept of isolated, open and closed systems, with hierarchical nested properties, as a methodological device to unify physical, natural and human sciences. In his discussion, Chorley went far further than Strahler (1952) in arguing for the complete abandonment of Davisian, historical approaches. This message was reinforced by two conferences at Madingley Hall outside Cambridge organised by Chorley and the human geographer, Peter Haggett, in 1963 and 1966. These really marked the onset of 'an advertising campaign' for the quantitative or reductionist approach (Chorley and Haggett, 1965, 1967). They were followed by two major discussions of the systems approach: *Physical Geography: a Systems Approach* in 1971 of which I was co-author; and *Environmental Systems: Philosophy, Analysis and Control*, 1978, with R.J. Bennett. Neither was really a success (see Kennedy, 1979).

What was far more important in my view was Chorley's major collaboration with his former Oxford Tutor, R.P. Beckinsale, on the first three volumes of *The History of the Study of Landforms* (1964, 1973, 1991). The first of these, in particular, although undoubtedly over Whiggish in its vision of the inevitable dawning of the Quantitative, Process-dominated day, was, nevertheless a genuine eye-opener. It is also a jolly good read. (The fourth volume, currently in preparation, is being edited by T.P. Burt of Durham University, one of Chorley's students, with others.)

And that, finally, is Chorley's key legacy a truly remarkable number of students who have stayed in academic geomorphology. Some of their names and their interlocking academic relations are indicated in Figure 8.2.

So Strahler's School has had a very deep impact on the Anglo-American academic community. Its final influence is before you. As I noted at the outset, Strahler's graduate seminars required that students study key literature and present abstracts of the works. This tradition descended to both Cambridge, with Chorley, and University of British Columbia, with Melton.

I did my undergraduate and doctoral work at Cambridge, with Chorley; and my Master's at University of British Columbia with Melton. So you are warned that the views expressed above are both literally and metaphorically partial.

Despite the widespreading diaspora of 'School of Strahler' products, it is notable that Strahler's original emphasis on attempts to treat morphological elements of landforms statistically has really not proved durable (although see Schumm, 1997). Instead, there has been a general move towards mathematical and physical treatments which concentrate on processes, often on very small temporal and physical scales. Two of those who feature in Figure 8.2 – Kirkby and Dunne – in fact were perhaps as strongly influenced by the other major American geomorphologist of the latter part of the twentieth century, Leopold, whose approach was in many ways different to that of Strahler.

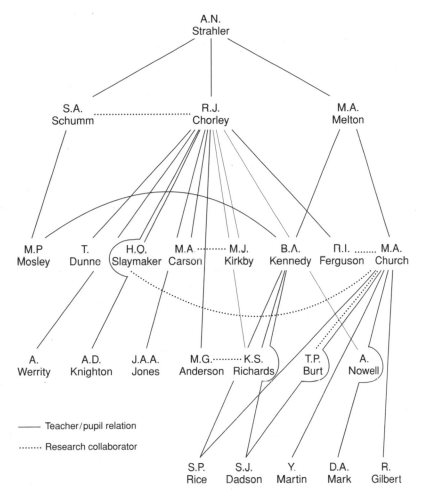

Figure 8.2 *Part of the 'family tree' of A.N. Strahler and his pupils.*

LUNA B. LEOPOLD AND HIS COLLEAGUES

A tremendous contrast to Strahler's personality, approach and basically academic background is provided by Luna B. Leopold (1915–), a son of the important American conservationist, Aldo Leopold (see Carter, 1980). Trained as a civil engineer, after a PhD at Harvard, he joined the U.S. Geological Survey and became its chief hydrologist. (He later moved to Berkeley.)

Leopold's crucial contribution was to bring into geomorphological focus the basic engineering concepts describing river behaviour (many of them developed in the eighteenth century). These he coupled with field observations and turned into the concept of 'hydraulic geometry' (but see Mackin, 1963). Two key papers are: Leopold and Maddock, 1953, 'The hydraulic geometry of stream channels and some physiographic implications' where the basic concepts of fluvial engineers were introduced and Leopold and Wolman, 1957 'River channel patterns – braided meandering and straight'. The latter demonstrated that it was possible to discriminate between braided and meandering channels on the basis of their discharge and their slope. For a given discharge, braiding occurs at a higher slope; for a given slope, at a higher discharge. This seemed to make almost magical sense of the bewildering sequence of river plan forms.

Undoubtedly Leopold's most lasting legacy will be the 1964 textbook *'Fluvial Processes in Geomorphology'*, which he co-authored with M. Gordon Wolman of Johns Hopkins and John P. Miller of Harvard. (Most unfortunately, Miller died in 1961 from bubonic plague contracted doing fieldwork in New Mexico.) This work has enormous verve and appears to reduce the chaos of fluvial landscapes to clear physical principles. It is, however, predominantly a work about the semi-arid American South West – although Wolman's fieldwork was in Maryland – and it tends to assume that *all* channels are alluvial and therefore mobile.

Amongst Leopold's key collaborators have been T. Maddock Jr., W.B. Langbein, M.G. Wolman and T. Dunne (originally one of Chorley's students). Leopold's influence has been very important and his approach has certainly come to dominate in studies of rivers. Despite some ventures into broader discussions (notably and, I feel, not very successfully, his 1994 *A View of the River*), Leopold's generally rather narrow, engineering-based approach has fitted late twentieth century concerns more closely than Strahler's broader-based work. Despite the lack of many direct academic descendents, Leopold's view of the river and the fluvial landscape has largely prevailed.

OTHER KEY WORKS AND WORKERS

The reductionist movement had its greatest success in Anglo-America. M.J. Kirkby, at Bristol and then Leeds, has devoted almost 40 years to building a model of slope evolution based on rather limited field evidence and the Continuity Equations which is, in its way, just as elaborate, didactic and theoretical as the Davisian cycle. M.A. Church, at University of British Columbia, has focused on river processes: some of his key students are D.A. Mark, R. Gilbert, Y. Martin, S. Rice and S. Dadson.

Wolman and Miller produced a very important work in 1960, trying to evaluate the significance in process terms of events of variable magnitude and frequency. The initial view that 'medium' events with an 'intermediate' return period – some 2/3 years – were crucial was very much conditioned by the semi-arid, mobile channel studies of Leopold and others. A later version (Wolman and Gerson, 1978) allowed for a consideration of bedrock cases (recall Gilbert's classification, Figure 6.5) and opted for somewhat larger and more widely spaced events. It was still unclear, however, how to reconcile the occasional massive landslide (or flood, in a bedrock channel) with the outcome of events which simply shift loose debris or soil.

A major study which addressed that point, in Swedish Lappland, was Anders Rapp's extremely important investigation (1960) of the rôle of different slope and stream processes in a subarctic catchment. This painstakingly distinguishes between the work done in changing hillside forms – where rock avalanches predominate – and the work associated with the transport of sediment out of the basin, where solution in the rivers is the major influence. This remains (and see Rapp, 1985, for an update) one of the finest quantitative process studies in existence.

Sweden had produced several earlier classic studies of fluvial processes: by F. Hjulström (1935) on the River Fyris and by A. Sundborg, on the Klarälven (1956); see Beckinsale and Chorley (1991, p. 182).

Mention has been made of Adrian Scheidegger's *Theoretical Geomorphology* (1961); on the whole, however, the German-speaking world was rather slow to follow the Anglo-American lead. An exception was Frank Ahnert, who had taught at the University of Maryland and who became particularly convinced of the importance of modelling slope development (see Ahnert, 1987).

Major studies on, essentially, the engineering properties of hillsides, came from M.J. Selby, an Oxford graduate established at the University of Waikato in New Zealand from the 1960s (see Selby, 1993) to the present.

One could go on. But what is worth noting is that the whole thrust of this emphasis on 'the dynamic basis of geomorphology' really was predicated upon the ability to make rather simple connections between process (or force) and form (or resistance), whether in mechanical or thermodynamic terms. It was such connections which could be described by regression equations, especially log-log ones. And, if one was working on the abbreviated and fast-abbreviating time-scale of doctoral studies or grants, one tended to be working with forms which were relatively mobile. From this and from a harking-back to an imperfect memory of G.K. Gilbert, came a vision of landscapes where processes were striving, ineluctably, to create mathematically describable equilibrium forms (see Thorn and Welford, 1994 and discussion). This ignores two crucial things.

First, that action and reaction are *only* equal and opposite in very, very simple systems – largely chemical ones. Resistance is not generally a linear function of applied force.

Second, virtually all forms are legacies of past applied forces. It follows that the sequence of events is often – probably always – a crucial determinant of landform and landscape response.

Unfortunately, these two drawbacks present very real analytic difficulties. Just as fluvial geomorphology was driving firmly down the path of simple Newtonian mechanics and the search for equilibrium forms, other parts of the scientific community were waking up to the crucial nature of history and sequence. It took a theoretical chemist – the Belgian, Ilya Prigogine (1917–2003) – to bring this to the fore, in 1977.

Chapter 9
Reinventing the Earth, 1977– : *Homo sapiens,* History and Microorganisms

There are more things in Heaven and Earth, Horatio,
Than are dreamt of in your philosophy.
(*Hamlet*, I, v)

INTRODUCTION

The last quarter of the twentieth century seems to me to mark important shifts in the way geomorphologists regard earth surface processes.

We may take Schumm's 1977 work *The Fluvial System* as a distillation of the 'family tree' from Davis, through D.W. Johnson to Strahler's revolutionary reductionist approach. At about that time it is reasonable to consider that Western humanity's view of the relationship between *Homo sapiens* and Nature – which has always been paradoxical (see Chapter 2) – came to take on (or invent) its current and profoundly schizophrenic character. It was in 1977 that Ilya Prigogine delivered his Nobel lecture (published 1978), in which he insisted on the significance of irreversible change – i.e. history – even in physics and chemistry. Also in 1977, the team of the submersible *Alvin* discovered the extraordinary and totally unexpected chemosynthetic ecosystems at hydrothermal vents 2 km below the sea surface on the Galápagos rift system (Jannash and Mottl, 1985; van Dover, 2000). And finally, in 1979, James Lovelock published *Gaia*, his startling vision of Earth as controlled by organic rather than inorganic processes.

This chapter will begin by an examination of the 'standard' view of fluvial geomorphology, encapsulated by Schumm's work, and the 'mainstream' developments of the following quarter century, especially in terms of the extension into 'new' environments, notably Australia, South Africa and Antarctica.

Second, taking the increasing association of geomorphology with engineering solutions, I shall try and consider the curious fashion in which humanity's contemporary vision – which is in many ways a re-invention of Dana's 'A world for mind' – produces schizophrenic responses to the variable magnitude and frequency of natural events.

Third, many geomorphologists – as we have seen in Chapter 8 – have been resolutely pursuing increasingly reductionist approaches to the study of earth surface processes, which, of course, chimes neatly with the necessarily short-term emphasis of both engineers and politicians. This emphasis has been justified, theoretically, by recourse to the supposed ahistorical stance of physicists and may be traced to the views of Strahler (1952) and Chorley (1962). The problems of sequence and of spatial scale which were supposedly to be overcome by the use of sophisticated computer modelling and General Systems Theory somehow did not succeed (see Kennedy, 1979). Hence the importance of Prigogine's explicit (though not unique) recognition that even the 'pure' sciences of physics and chemistry needed to grapple with questions of historical contingency. This has taken some geomorphological studies into the field of non-linear dynamics, with consequences which are as yet unclear, but which may produce the invention of an approach that can deal with both the immanent and the contingent (Simpson, 1963) in earth surface processes.

Finally, our vision of the Earth, ever since the mid-eighteenth century, has been one dominated by inorganic forces which are working to some degree to the benefit of *Homo sapiens*. The *Alvin* discovery has led to the opening of an enormous Pandora's Box of extremophile microorganisms (see Gross, 2001) of immensely antique lineages, whose rôle in earth processes covering three-quarters of Earth's history is only beginning to be guessed at, providing yet another demotion of Humankind as true Lords (and Ladies) of Creation. This may represent the most dramatic reinvention of the Earth since Hutton and may be linked to Lovelock's *Gaian* visions (1979, 1988, 1995).

THE FLUVIAL SYSTEM

As I noted in Chapter 8, Stan Schumm is, of Strahler's major pupils, the most genuinely successful and internationally influential as a geomorphologist. Three of his works have had very widespread attention. The 1965 paper (with Lichty) entitled 'Time, space and causality' attempted to wrestle with the enduring dilemma of scale dependence in earth surface phenomena: how can one link apparently time-invariant behaviour on the small scale of (say) a river channel's bedforms with the progressive changes of landscapes over geological time? Schumm and Lichty's attempt at reconciling these scales was not entirely successful (see Kennedy, 1997b), but it has been a durable reminder of basic problems to which we shall return below.

The 1965 paper features in Schumm's two books: *The Fluvial System* (1977); and *To Interpret the Earth: Ten Ways to be Wrong* (1991). It is on the former that I shall focus, since – unlike many texts – it is a personal and hence necessarily idiosyncratic view, rather than a compendium of other people's ideas.

The first wholesale discussion of exclusively fluvial processes since (possibly) Greenwood, 1857, came – not surprisingly (see Chapters 6 and 7) – from North America, with Leopold, Wolman and Miller's revolutionary *Fluvial Processes in Geomorphology* (1964). This was not only firmly rooted in engineering-scale investigations, but dealt virtually exclusively with North American examples.

(An English version, with its own parochial emphasis, was Gregory and Walling's 1973, *Drainage Basin Form and Process*.) But by the 1970s the great international movements of scientists including earth scientists were well underway. Schumm (1977) therefore contains important sections on Australian rivers, representing his own field studies. This widening of geomorphological experience has continued and increased (cf. Knighton, 1998) and is perhaps best demonstrated by the truly global range of examples covered in Miller and Gupta's edited 1996 *Varieties of Fluvial Form*, where examples range from Antarctica to India, via South Africa, Australia and New Zealand. The questions (and possible answers) posed by different parts of the Earth's surface *do* vary. Both the very old (cf. Ollier, 1991) and the geologically very new (cf. Selby, 1985) as represented by much of Australia and of New Zealand respectively, present startling contrasts to both Europe and the American Southwest, from which much of the received wisdom of the mid-twentieth century fluvial geomorphology derived.

A second important feature of Schumm's work is the combination of carefully designed field investigation with laboratory experiments. The use of a huge 'sand box' allowed Schumm and his students to replicate drainage network developments in ways which not infrequently led to surprising findings. One major concept which emerged was that of 'complex response', with respect to a change in base level. For a century or so, it had been assumed that pairs of terraces along river courses each represented a single drop in local base level. Schumm and Parker, however, found, that when they created an artificial fall in base level, *two* pairs of terraces resulted from the cut-and-fill processes of the headward promulgation of the wave of erosion (see Schumm, 1977, figure 4-12). In the twenty-first century, such hardware experiments have largely been replaced by computer modelling (cf.. Rodriguez-Iturbe and Rinaldo (1997), and see the penultimate section of this chapter), yet Schumm's experimental approach in the tradition of Hutton, Darwin and Gilbert, has proved extraordinarily fruitful of interesting questions to ask of real landscapes.

Allied to the notion of complex response in Schumm's work is that of the significance of *thresholds*, both extrinsic (e.g. when climatic variables exceed some limit) and geomorphic (e.g. when a flood overtops a river's bank). The importance of thresholds (see Coates and Vitek, 1980) is that they represent departures from the straight Newtonian concept of a simple connection between applied force and geomorphic response. An early example came from Schumm and Chorley's (1964) reconstruction of the fall of Threatening Rock, a massive monolith of sandstone whose collapse, in 1941, they considered was the product of small, incremental displacements due to frost and rain, the last one of which triggered the collapse: any system with multiple embedded thresholds creates extreme difficulties for simple relations between cause and effect (see the penultimate section of this chapter).

Finally, although both *The Fluvial System* and *To Interpret the Earth* have an underlying recognition of geological-scale events, the former is also very much attuned to the human timeframe and, in particular, the notion of 'large scale' events and their impact, transient or long-lasting, on the landscape.

All in all, Schumm's 1977 work represents to my mind not only a very fruitful encapsulation of mid-twentieth century approaches to the study of fluvial forms and processes, but also adumbrates two aspects of the study of geomorphology which have become more evident in the last quarter century. The first is, of course, that of 'the human impact'.

'A WORLD FOR MIND'

This was you may recall, Dana's vision of the Divine purpose which directed the operation of geomorphological processes. As Davies (1968) and Gould (1987) both emphasize, humanity – or, at any event, Western, educated humanity – has had a long and troubled experience of attempting to reconcile observations of earth surface processes with what are felt to be the necessities of the Almighty's designs for human survival (or destruction). Lyell's *Principles* in its various editions (see Kennedy, 2001) had as a major objective the demonstration that processes observable by contemporary nineteenth century society had sufficient magnitude and efficacy to actively create and modify global topography without Divine intervention. As the twentieth century progressed, however, this vision has become curiously subordinated to an excessive 'environmentalist' view whereby the rapid growth of the global population (6500 million or so, and rising) of that medium-sized omnivore *Homo sapiens sapiens* has replaced the Almighty as omnipotent and omniscient. In this view, all so-called 'natural disasters' from enhanced global warming, through sea-level rise, to tsunamis, hurricanes, floods and influenza, are now to be blamed on humanity's malfeasances (although see Lomborg (2001) for a counter view) and, in consequence, must necessarily be curable by human counter-actions. I shall try to illustrate some of the painful (and to my mind both absurd and dangerous) lapses in rational thinking which result in the present widespread belief that Nature is both essentially benign and necessarily subservient to humanity's needs. (See below for the Boxing Day tsunami of 2004.)

There seem to me to be three major reasons for the view of humankind 'bestriding this narrow world like a Colossus' which has grown so dramatically in the Western world since the last quarter of the twentieth century, to the extent that the geological era in which we live is, by some, no longer termed the Holocene, but rather the Anthropocene (Crutzen and Stoermer, 2000), or the time in which human influence on the Earth has become as extreme as that of all natural forces (Finningan, personal communication, 2004).

First, we happen to have lived through more than 125 years without any major demonstration of the power of natural forces having impinged upon the collective consciousness of Western Society. The last one (before 26 December 2004) I would argue, was provided by the cataclysmic eruption of Krakatau in the Malay Archipelago in 1883 (see Thornton, 1996; Winchester, 2003). This, through its immediate repercussions in South East Asia but, more important, by the global impact in terms of meteorological and climatic effects and the truly widespread influence upon sea level – both the dramatic and deadly local tsunami and the tiny yet mind-boggling impact on European tide gauges – genuinely shook the world.

What has there been of comparable impact since? The San Francisco Earthquake in 1906? the Bangladesh floods of 1970? the eruption of Mount St Helens, 1980? or of Mount Pinatubo, 1991? the Kobe earthquake, 1995? Well, each and all of these have had terrible local consequences, but I doubt if they have left any lasting impression on the global (or Western) mind.

An interesting series of comparisons, if you wish to assess how hazards are viewed, is to consider the following: Chernobyl (1986), Bhopal (1984) and Lake Nyos, Cameroon (1986). The first, of course, not only involved that bogey of the Chattering Classes, nuclear power, but had direct economic repercussions across north and west Europe; the second, although similarly the product of human error, not only involved 'merely' a petrochemical plant that ran amok, but virtually all victims – many of whom suffered appalling long-term effects with little or no treatment or compensation – were Indian nationals: 'out of sight, out of mind'?; and Lake Nyos? Did you miss that altogether? It seems a blanket of carbon dioxide gas was released by the natural overturning process in a volcanic lake, which suffocated all 1700 or so people (and other animals) in the vicinity. But that was in Africa, of course . . . and as far as I know no-one has yet found a way to invoke a human agency . . .

However, both this highly partial view of technological traumas and the frankly smug and purblind view of Nature and its workings are predicated upon a very, very partial and ahistorical vision of even 'recent' Earth history: say the last 10,000 years or so of the Holocene Epoch (i.e. after the retreat of the last major ice sheets). We find it easy to forget that largeish parts of the world had few or no inhabitants for much of that period. The great caldera (collapsed volcanic crater) represented by Lake Taupo in New Zealand was formed by a truly enormous eruption in about AD 186, well before the arrival of the first Maoris. Even the 1811–12 New Madrid earthquake swarms of the central Mississippi Valley, which – for a time – reversed the river's flow, occurred far from telegraphs let alone helicopters and camcorders.

Large natural events *are* generally rare on the geological timescale, but they have not ceased (see below).

The second factor which, in my view, has led many of us (and all politicians) to underestimate the power of Nature is the ever-shrinking timeframe of earthsurface process studies (contrast with Ollier, 1991). Early work (e.g. by Saarinen, 1966) on hazard perception demonstrated that anything which recurs at longer intervals than 4 or 5 years is either forgotten or downplayed: this is, of course, especially likely in highly mobile societies like that of the USA, where there are relatively few 'oldest inhabitants' to provide a local race memory.

We think that each natural phenomenon exhibits what we term an 'event series'. That is, if the general controlling conditions remain constant, there is a more-or-less definable relation between the magnitude of happenings and their frequency of occurrence. If we rank the size of (say) river floods over a period of record, then, making use of a statistical device associated with E.J. Gumbel (see Dunne and Leopold, 1978, chapter 2) we can attach a probable return period or recurrence interval to events of given size. So we can talk of the '10 year' or '50 year' flood. By this we mean that, if the controlling conditions remain the same, there is the expectation of one such sized event to occur at random within any particular 10 or 50 year period on average. As we noted in Chapter 8, a very influential paper by Wolman and Miller (1960) suggested that the key formative events in many fluvial landscapes occurred about once every 2–3 years. Wolman and Gerson (1978) revised that estimate to about 10 years. Either seems, frankly, to provide a very false sense of reality in a world almost 4600 million years old.

However, there are two, real difficulties with these recurrence interval concepts: we generally dispose of very, very short direct records of events; *and* it is extremely unlikely, as the length of record increases, that the controlling conditions remain the same. One of the longest records, that of some 1000 years of the annual Nile flood, in Lower Egypt (Hurst, 1950), was rendered of limited modern value by the building of the two Aswan Dams in 1902 and 1970.

What records do we have? Well, as we noted in Chapter 5, Darwin's earthworm experiment extended over 30 years. Even more remarkable (see Johnston, 1994) is the series of Rothamsted experiments established by Lawes and Gilbert, from 1843, some of which continue to this day. Parts of the Broadbalk field at Rothamsted have been continuously cropped to winter wheat for more than 150 years, with no addition of manure. The graph of yield (Johnston, 1994, p. 32) therefore must represent net *erosion* of the field's soil by the annual harvesting and removal of the crop's biomass (see Kennedy, 2000). But, evidently, such long-running experiments are both few in number (although see Leigh and Johnston (1994) for a fascinating account) and relate to low intensity processes. (I cannot but wonder how far the susceptibility to water erosion of the long-cropped, shallow soils of the South Downs, as documented by Boardman (e.g. 2003) may not link to the kind of removal of mass to be noted at Rothamsted.)

What can be done to extend records? There are various ways in which dating of either deposits – (including stalactites and stalagmites, termed speleothems) or eroded surfaces (or, increasingly, both; cf. Dadson *et al.*, 2003) may be linked to estimates of the magnitude of events, so that a record of river floods (in particular) may be extended back beyond the period of observation. Baker *et al.* (1988) provide a wide-ranging discussion of ways of extending the event series of river floods. Reusser *et al.* (2004) have used the amount of cosmogenic beryllium (^{10}Be) present in surface samples to evaluate a 35,000 year history of the incision of the Susquehanna and Potomac rivers into bedrock. Dadson *et al.* (2003), examining erosion rates in Taiwan and their dependence on uplift and precipitation (notably from typhoons), use a range of techniques, including radiocarbon dating of Holocene river terraces and apatite fission-track dating of exhumed surfaces. This latter is a complex, new technique (see Dadson *et al.*, 2003, p. 651) which allows estimates of the time at which buried rocks from 3 to 5 km depth have been exposed at the surface, using the tracks made by the fission of the mineral apatite. The technique permits estimates of age back to a million years.

Nevertheless, the reconstruction of all event series hinges crucially on the stability of the environment. Although it may be reasonable to use the concept of a '50 year rainstorm' or 'tidal surge' in southern Britain, there will be rather few drainage basins where by dint of land use changes and, in particular, the encroachment of building on floodplains, the flood event series is genuinely the same as that of 50 or 100 years ago.

All of which tends to the view that Nature is becoming more inimical to the happy existence of suburban householders and that Something Should be Done About it . . . Because the third factor at work is that, with some 6500 million human beings now clinging to the surface of the planet, extreme events – whether as natural as a meteorite strike or as human-induced as the inevitable downstream flooding which follows upstream forest clearance – cause increasing problems to increasing numbers of *us* (see Diamond, 2005). And there can be no doubt that the Chattering Classes have justification in their belief that *H. sapiens* is now and probably always has been, over the past 100,000 years, an extremely effective agent of surface alteration (albeit without the omnipotence now often assumed). Forget those ideas of Rousseau's Noble Savage in tune with the rest of Nature. Any group of moderately large omnivores, especially ones able to manipulate fire, is going to be able to change the environment to suit itself and, in so doing, alter the way in which geomorphic processes operate. Today, the term Anthropocene would, indeed, seem fitting if we consider Hooke's (1999) data (Figure 9.1) for the balance of respectively, 'natural, fluvial' and 'anthropogenic' erosion in the USA . . .

I am trying to highlight a major paradox in the view of the relation between Mankind and Nature held now by many people in the First World and which amounts to a true, new Ruling Theory or paradigm. Nature is seen as fundamentally beneficent and there to serve humankind's needs. When Nature appears malign, it is assumed that this results from some failure in human control as humankind is seen as infinitely capable of transforming and controlling nature, given sufficient capital and good will. This attitude is well encapsulated by the former President of Bangladesh who called upon the rest of the world to halt the annual flooding which afflicts his country . . . Did he wish the Monsoon circulation to be deflected? The Himalayas not merely reforested, but flattened? The topography of the Bay of Bengal to be modified? And the atmospheric circulation which produces typhoons in the Bay of Bengal to be deflected, too?

There is, it appears, a great truth about attempts to control Nature for human ends, which, as Needham (1971) explains, the Chinese had grasped two millennia ago. The greater the control, the more the immediate economic advantage, but the greater the ultimate environmental disaster. Put the river behind flood walls and levies and build all over the flood plain, but do not be surprised when the river escapes and wreaks havoc. The alternative, of letting the river roam across the floodplain, at the cost of some lost lives and major lost economic opportunity, becomes less and less acceptable as the pressure of population upon flat and potentially arable land grows.

We are, then, very sure that we are the apogee of Creation, that the Earth is *ours* and that our technology can counter any Earth surface processes which

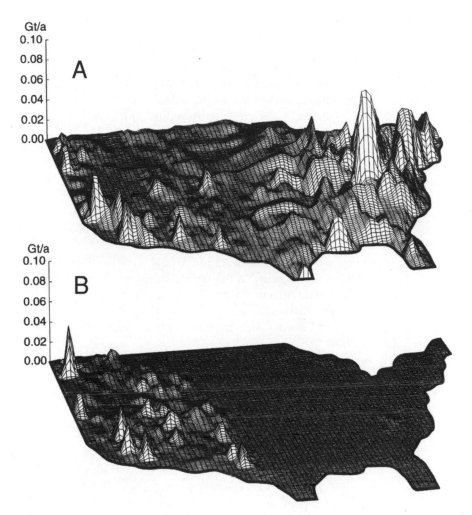

Figure 9.1 *Comparison of human (A) and natural (B) contributions (in gigatonnes per annum) to earth movement. (From Hooke, 1999, figure 1.)*

threaten to inconvenience us. As the memory of really extreme and genuinely natural events fades, so humanity assumes that it is some failure in the *human* system which must, *inevitably*, be the *sole* cause of any environmental hazard.

The preceding paragraphs were written before the Magnitude 9 earthquake which occurred on 26 December 2004 along the major subduction zone to the west of Sumatra. Here a massive section of the ocean floor contact between the Indian and Eurasian plates was disrupted. The resulting tsunami was responsible for at least a quarter of a million deaths around the Indian Ocean, mainly in western Sumatra, but extending as far as Somalia and Kenya to the west and Myanmar to the north.

Winchester (2003) considered the Krakatau eruption and tsunami of 1883 to be the first 'global' disaster. Some repercussions of the explosive collapse of the volcano (like the airborne shock waves) were actually noted all around the world; but the telegraph system also permitted news of the event to travel almost instantaneously. The 2004 event was even more global, in that images and accounts were available on televisions around the world within a matter of hours. (The inhabitants of fishing villages in eastern Sumatra did not, however, learn of the event for three days: M.E.A. North, personal communication.)

What is of interest here were the reactions. All initial interviews with survivors registered awed shock at the sheer force of the water, beyond anything they could imagine or conceive possible. Within a week, the tone from outside the disaster area was shifting towards: 'This (or some of this) was or should have been preventable: it should never happen again.' It is now proposed to install a tsunami warning network in the Indian Ocean, to match that in the Pacific, even though tsunamis are very rare in the former body of water.

You might ask yourself if you think reactions would have been similar if the event had affected the South Atlantic? If the waves – generated from submarine events on the southern Mid-Atlantic Ridge – had smashed into the shores of Argentine Patagonia, the Falklands and Tristan da Cunha? I will admit that impacts at Cape Town and Buenos Aires might have been registered on CNN or the BBC World Service.

My point is simple: the impact of the December 2004 event was primarily viewed through the experiences of thousands of First World tourists. Millions of those who were not on holiday on 26 December 2004 could, nevertheless, envisage the beaches and resorts (and local communities) they knew from previous visits . . . There will be outrage, rather than awe or shock, should a comparable event occur in the next year or two.

There is no doubt that a key element in our vision of *Homo dictator* has been the staggering general success of Western engineering technology, with its necessary emphases upon simple relations between action and reaction. That, in turn, emphasizes the human temporal and spatial scale, its inevitable intersection with political visions and its inherently reductionist principles. However, the latter part of the twentieth century has produced a number of counter ways of regarding and treating mathematically change over time and space, one of the most significant statements of which, to my mind, being given by the Nobel Laureate, Ilya Prigogine, in his Nobel speech of 1977.

EQUILIBRIUM, HYSTERESIS OR HISTORY? CAN WE INVENT NEW MATHEMATICAL TOOLS TO COMBINE THE IMMANENT AND CONTINGENT?

Figure 9.2 comes from Prigogine, 1978. It shows the nature of solution activity over time. Up to point A, a reaction may be thought of as not only completely reversible, but also ahistorical. Consider the phase changes that water undergoes between ice, liquid and gas: these changes are completely reversible and, in pure

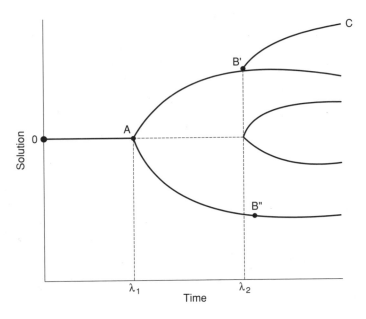

Figure 9.2   *Time and change. (After Prigogine, 1978, figure 4.)*

water, leave no trace. However, once a reaction passes the first bifurcation point, A, there are two, possible states which correspond to the same length of reaction time. We cannot now *a piori* determine whether we shall get to B' or B". It may be possible to revert to A or even O, but the reversion will be historically explicit. This corresponds, in my view, to the well-known phenomenon of hysteresis.

Figure 9.3 illustrates this with respect to the relationship between the amount of water (Q) and sediment (Qsed) in a river. Figure 9.3A shows an apparently simple relationship: as force (Q) increases, so the amount of work done (Qsed) increases, too. As discharge falls sediment transport falls. But, in reality, the relation is time and space specific. Figure 9.3B illustrates a case where there is massive movement of sediment as discharge rises, exhausting the available supply, so that there is little or no transport as discharge falls. Figure 9.3C shows the reverse case. The values of Q(sed) linked to Qx are, evidently, no simple function of the value of discharge. The apparently atemporal relationship of A in Figure 9.3 is shown to be historically contingent. Now consider what is happening by the time we pass the second bifurcation in Figure 9.2: there is a *sequence* of specific thresholds passed and the reaction is now most definitely one with an irreversible component, or history. In fluvial terms, think of the cut-off of a river meander: did the cutting of the new channel happen in one large flood, by the process of avulsion (cf. the channel changes caused by the Lynmouth flood of 1952, documented by Anderson and Calver, 1977) or, rather by 'a last straw that broke the camel's back' process where a narrow neck of land was finally undermined? Whichever the historical reality, we have a spatially and temporally specific change which is now irreversible.

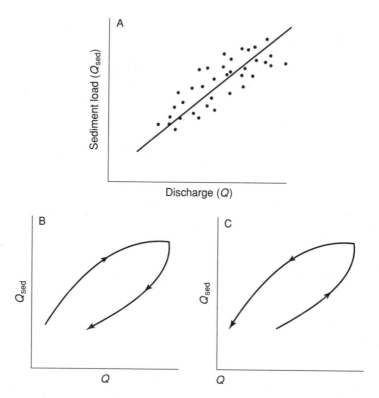

Figure 9.3 *Hysteresis effects concealed in an apparently straightforward relation between discharge (Q) and sediment load (Qsed). For explanation, see text.*

Prigogine describes the situation thus: '[It is interesting that] bifurcation introduces, in a sense, "history" into physics. Suppose that observation shows us that the system whose bifurcation diagram is represented by Figure 9.2 is in state C and came there through an increase in the value of λ. The interpretation of this state . . . implies the knowledge of the prior history of the system, which had to go through the bifurcation points A and B. In this way we introduce in physics and chemistry a historical element, which until now seemed to be reserved only for sciences dealing with biological, social and cultural phenomena' (1978, p. 781). He might have added: geological and geomorphological.

This very important shift in the view of the significance of contingency and hence, sequence, was not unique to Prigogine. As the account by Gleick (1987) makes plain, an earlier – and now often misquoted – vision of the rôle of contingent circumstances was put forward in the early 1960s by the meteorologist Edward Lorenz. His 'Butterfly Effect' – the notion that a butterfly stirring the air today in Peking can transform storm systems next month in New York (Gleick, 1987, p. 8) is better termed 'sensitive dependence upon initial conditions'. That is, the precise initial configuration upon which the forces of physics and chemistry go to work has an acute and unpredictable control over the resulting outcome.

The mixture of the deterministic (described by the laws of physics and chemistry: the immanent or abiding process of Simpson's (1963) discussion) and the probabilistic element introduced by the vagaries of temporal and spatial configurations (Simpson's historically contingent notion), has now grown into the ever-ramifying concepts broadly termed non-linear dynamics (see Hall, 1992; Turcotte, 1997). These include Catastrophe Theory, sandpile models, cellular automata models and a whole range of ideas termed self-organized criticality (see Bak, 1996).

One area of especial interest to students of the Earth's surface came from the work by Benoît Mandlebrot on the mathematical device known as the fractal (see Gleick, 1987, pp. 83–118). This is essentially a measure of the complexity of an infinitely repeating pattern, which is scale free and, therefore, suggests no detectable historical legacy in the phenomenon concerned. This is, it would seem, very important. If we can *eliminate* spatially – or historically – contingent variations, then our explanations or predictions should become both simpler and more successful. A direct geomorphological example was provided in Mandelbrot's article 'How long is the coast of Britain?' (1967). The answer turns out to be dependent upon the fineness of the measurement device (if you measure in metres, the coastline is much longer than if you take 1 km lengths). But, the trace of a section of coast measured at different scales looks the same. There is an intrinsic fractal dimension (in the case of the coastline it seems to be ≈ 1.05: Rodriguez-Iturbe and Rinaldo, 1997, p. 112): put another way, views of the coastline at *any* scale exhibit what is termed self-similarity. The shapes are scale free: this is also true of the cross-sectional shape of glacial troughs, but not of river channels (see Kennedy, 1984); and of drainage densities (see Church and Mark, 1980). However, the explicit application of these ideas and other ideas of non-linearity in geomorphology began rather belatedly (cf. Phillips, 1992), and their consequences are still unclear.

There seem to be two areas of particular interest. First, how far can this family of new ideas help resolve persistent problems with 'regular' spatial patterns – such as river meanders – which are never as regular as they seem (cf. Stølum, 1996)? Second, can they help account for the irregular temporal distribution of discrete process events, such as landslides (see Malamud et al., 2004)?

But it is too soon to know how far these new ideas will increase our understanding of earth surface processes and landforms. They are certainly taking us away from a simple Newtonian or even equilibrium vision, and an area which is attracting particular interest is the possibility of developing predictions for the occurrence of those 'high magnitude/low frequency' events which are termed natural hazards or natural disasters by human societies (Malamud et al., 2004). As many of these involve the crossing of intrinsic or extrinsic thresholds, in historically and spatially defined sequences, the development of mathematical treatments capable of explaining or even predicting events would represent a significant new invention.

The changes in technology and in approaches to 'extreme' events outlined in this section, as well as the increasingly successful attempts to extend dating techniques to longer time periods and new earth materials, are, I think, having some impacts on our professional view of earth surface processes. But I think we have yet to come to terms with the full implications of the discovery of the hydrothermal vent communities in 1977. This implies that the Earth dances, perhaps, to a very different drummer not merely than *H. sapiens sapiens*, but – more profoundly – than the suite of largely mechanical endogenetic and exogenetic forces we have assumed as dominant, ever since Buffon and Hutton

THE TREE OF LIFE

Most biology textbooks have – or had – a diagram which shows the Kingdoms of macroscopic organisms (the eukaryotes) – fungi, plants and animals – as the dominant life forms of the planet.

Figure 9.4, on the other hand, is a more modern view. The eukaryotes are seen to be a minor division of living things. Gross (2001, p. 132) suggests that bacteria, who dominate Figure 9.4 have been with us for some 3800 million years (out of the 4550 million we believe is the total). And there was nothing else until the development of algae about 1700 million years ago.

Why should this be relevant to an account of earth surface processes? Well, after the discovery of the Galápagos rift vent ecosystems, it became clear that the rules of life were not as we had imagined them and must be reinvented. Bacteria and other microorganisms of all kinds turned up in all kinds of improbable places: not merely coping with temperatures of 110°C near the 'black smoker' towers of precipitates at the ocean ridge vents, but at pressures of 1050 atmospheres from 10,500 m below the sea surface in the Marianas Trench; after vacuum drying for 6 weeks in extremely saline environments; at pHs of below 2.0 (stronger than dilute hydrochloric acid); and equally at a pH of more than 10.0 (approaching that of ammonia); or in the ice covering Lake Vostock in Antarctica (data from Gross, 2001). And they were, of course, busily *living*, not just in the ocean depths, but – and this is where they really should begin to interest earth scientists – deep in the Earth's rocks and pretty much everywhere at its surface. There is, too, the genuine prospect that bacteria or something like them will turn up on Mars and some of our other neighbours in the solar system. Wherever they are, or have been, they will – just like *H. sapiens sapiens* for the last twinkling of a geological eye – be actively converting their environment to one favourable to *their* existence. This is the truly revolutionary vision put forward by the independent scientist, James Lovelock, whose Gaia concept (1979, 1988, 1995) is of a planet whose inorganic characteristics (notably atmosphere and oceans) are manipulated by its organisms. The best analogy to my mind, is that of the termites' nest. A vast and elaborate inorganic structure precisely, if blindly, engineered to maintain an optimum home for the builders and occupiers.

Lovelock's views have many of the characteristics of those of previous revolutionaries, notably Hutton and Wegener. Lovelock is to one side of the mainstream of late twentieth century science. He has focused on anomalies – such as

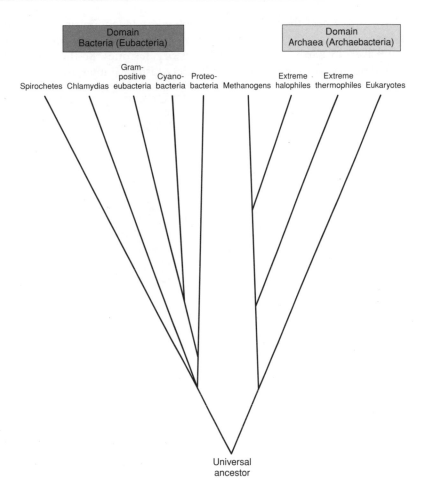

Figure 9.4 *A modern version of the 'Tree of Life'. (From Campbell, 1996, figure 23.)*

the oxygen content of the Earth's atmosphere – which are largely ignored by the dominant, inorganic Earth story. Finally, his vision coincided with the explosive discoveries of the ubiquity, variety and above all antiquity of microorganisms. Who knows how far we may have to revise geology to take account of this? We are already (cf. Viles, 1995) taking more account of the rôle of a range of micro-organisms (not just bacteria) in weathering processes and those processes, in turn, have important consequences for the levels of carbon dioxide in the atmosphere and, yet again, for global temperatures. Darwin's interest in the profound earth-forming effects of small organisms may have been truly prescient.

We have in prospect the need for the wholesale reinvention or refocusing of earth forming processes. Maybe rain and rivers will prove to have been the most minor of players in Earth – or planetary? – history/histories. So the story I have been unfolding may turn out to be one of purely historical interest!

Epilogue

This has not set out to be a Complete History of Ideas about the Earth. It is, rather, a notably partial and biased account, albeit one which has tried to link what I consider to be key aspects of the family history of modern geomorphologists (especially fluvial ones) to present and even future practice.

For those seeking more balanced – or possibly differently biased – accounts, I would recommend four works:

Walker and Grabau (1993) for a notable attempt to paint a picture of the development of geomorphology around the globe.

Rhoads and Thorn's (1996), edited account of the 27th Binghamton Symposium, devoted to the question of *The Scientific Nature of Geomorphology*.

Huggett (2003) as an up-to-date account of the prevailing Anglo-American view of geomorphology. It has the merit of giving explicit recognition to early ideas and it is also turning attention once more to the *historical* aspects of landforms.

Goudie's (2004) two-volume *Encyclopaedia of Geomorphology* produced by and for the International Association of Geomorphologists provides an up-to-date overview of all earth surface processes as we see them at the outset of the twenty-first century.

But, as a parting shot, may I suggest that we should all do well to recall W.M. Davis's 1926 exhortation to sit so lightly in thrall to our ruling hypotheses (or paradigms) that we feel no mental torment in having them refuted or overthrown. The last word in that respect must go to Mark Twain:

'In the space of one hundred and seventy-six years the Lower Mississippi has shortened itself two hundred and forty two miles. That is an average of a trifle over one mile and a third a year. Therefore, any calm person, who is not blind or idiotic, can see that in the Old Oölitic Silurian Period, just a million years ago next November, the Lower Mississippi was upwards of one million three hundred thousand miles long, and stuck out over the Gulf of Mexico like a fishing-rod. And by the same token any person can see that seven hundred and forty-two years from now the Lower Mississippi will be only a mile and three quarters long . . . There is something fascinating about science. One gets such wholesale returns of conjecture out of such a trifling investment of fact' (Mark Twain, 1883, *Life on the Mississippi*, chapter XVII).

Appendix I

AGASSIZ, J. Louis R. (1807–1873). Swiss expert on fossil fish, who promulgated (1839) the revolutionary concept of the Ice Age. Later Professor of Zoology at Harvard and firm opponent of Darwinian evolution.

BASSET, César-Auguste (1760–1828). French cleric, teacher and educationalist. Lived in exile during Revolutionary period. Translated (1815) Playfair's *Illustrations* into French and enthusiastically annotated the work with his own, supporting observations on British scenery.

BAULIG, Henri (1877–1962). French physical geographer, Professor, University of Strasbourg. Major supporter of Davisian views.

BECKINSALE, Robert P. (1908–1998). Oxford geographer, Tutor of Chorley and co-author of the first three volumes of *The History of the Study of Landforms* (1964, 1973, 1991).

BERTALANFFY, Ludwig von (1901–1972). Austrian born biologist. Professor, University of Edmonton. Formulator of General Systems Theory, 1950.

BOURGUET, Louis (1678–1742). French natural philosopher. Professor, University of Neuchâtel. Two key works (1729 and 1742) introduced the notion that the interlocking spurs (*angles saillans et rentrans*) of valleys demonstrated that they were erosional features (idea later challenged by Saussure).

BRETZ, J. Harlen (1882–1981). American geologist. Professor, University of Chicago. Correct suggestion that the Channelled Scablands of Washington and Oregon were the products of gigantic, glacial lakeburst floods was actively resisted by the USA geological establishment for 30 years.

BREISLAK, Scipione (1750–1826). Italian cleric and geologist. Early proponent of the equivalence between modern volcanic lavas and basalts. First drew attention to the significance of the Temple of Serapis, Pozzuoli as evidence for multiple historical changes in sea level.

BUCKLAND, William (1784–1856). English cleric and geologist. Reader in Mineralogy, Oxford University: later Dean of Westminster. His belief in the widespread evidence of Noah's Flood was overturned by Agassiz's vision of wholesale glaciation by land ice during the Ice Age.

BÜDEL, Julius (1903–1983). German climatic geomorphologist. Professor, Universities of Berlin and Göttingen. Major expeditions to Spitzbergen.

BUFFON, George-Louis L., Comte de (1707–1788). Great French naturalist and natural philosopher. Important geological/geomorphological ideas included: major extension of Earth history beyond the Biblical 6000 years, after physical experiments; the concept of successive 'Epochs' in Earth history, terminated by cataclysms; the first geomorphological map; the first numerical system of stream ordering.

CHAMBERLIN, Thomas C. (1843–1928). American geologist. His 1890 'Method of Multiple Working by Hypotheses' has been repeatedly reprinted and remains an excellent guide for the testing of ideas especially in historical science during the periods of 'normal' or 'puzzle-solving' science defined by Kuhn.

CHARPENTIER, Jean G.F. de (1786–1855). Mining engineer and pioneer Swiss glaciologist. Agassiz profited from his excursions with Charpentier to produce the Ice Age concept. Charpentier's views were subsequently eclipsed by Agassiz's fame.

CHORLEY, Richard J. (1927–2002). English geomorphologist, pupil of Beckinsale and Strahler. Professor at Cambridge University. Introduced Bertalanffy's ideas to geomorphology, 1962. With Haggett, launched the 'Quantitative Revolution' in British geography. Major work the co-authored *The History of the Study of Landforms* (1964, 1973, 1991).

CHURCH, Michael A. (1942–). Canadian fluvial geomorphologist, Professor, University of British Columbia. Only major student of Melton. Work is increasingly engaged with the application of physical principles to Earth surface processes.

COTTON, Charles (1885–1970). Major New Zealand geologist and geomorphologist. Professor of Geology, Victoria University of Wellington. Originally a strong adherent to ideas of Davisian cycles, his later work leant more towards views of the German school of climate geomorphology.

CROLL, James (1821–1890). Scottish scientist and proponent of the astronomical theory of variable solar radiation as a cause of glacial epochs. A notion vigorously disputed by Lyell, but later incorporated in Milankovitch's refined models.

CUVIER, Georges L.D., Baron (1769–1832). Great French naturalist and palaeontologist. Rejected 'outdated' Theories of the Earth, although he – of course – subsequently advanced his own versions (cf. 1825).

DANA, James D. (1813–1895). American geologist and zoologist. His participation in the 1838–1843 USA 'Exploring Expedition' confirmed his belief in the efficacy of modern fluvial processes. However, his religious beliefs led to a view of Earth history dominated by successive cataclysms, leading to the goal of 'a world for mind'.

DARWIN, Charles R. (1809–1882). Scientific genius whose early and late works were especially geological and geomorphological. His alternative theory of the origin of coral reefs (1842) proved far more fruitful than that of Lyell; and his study of earthworms (1881) foreshadows twentieth-century interests in the geological work of organisms.

DAVIES, Gordon L. (Herries) (1932–). Historian of Earth Science, Trinity College Dublin whose 1968 work forms a scholarly counterpoint to Chorley *et al.* (1964).

DAVIS, William Morris (1850–1934). American physical geographer, Professor at Harvard University. His invention of the pseudo-Darwinian concept of the Geographical Cycle in 1889 dominated much geomorphological activity well into the 1950s.

DE LA NOË, Gaston (1836–1902). French Lieutenant-Colonel, co-author (with De Margerie) of the first wholesale quantitative treatise on landforms (1888).

DE LUC, Jean A. (1727–1817). Swiss geologist, long settled in England. A violent and persistent critic of Hutton's views on the extent of geological time.

DE MARGERIE, Emmanuel J. (1862–1953). French geologist and co-author (with De la Noë) of the first wholesale quantative treatise on landforms (1888).

DURY, George H. (1916–1996). British fluvial geomorphologist, Professor at Sidney University and the University of Wisconsin at Madison. Took up Davis's work on underfit streams and devoted major effects to establishing relationships between underfit meanders and putative former bankfull discharges, resolutely repudiating the role of variable lithology.

DU TOIT, Alexis L. (1878–1948). South African geologist and early and strong adherent (1937) to Wegener's views on Continental Drift. The story of the conflicts aroused in USA geological circles by attempts to support his researches in eastern South America is related by Oreskes (1999).

DUTTON, Clarence E. (1841–1912). American geologist and seismologist. Member of Powell's team in the Colorado Plateaus. Important work on slope retreat and on isostasy.

ELIE DE BEAUMONT, J.-B.-A.-L.-L. (1798–1874). French geologist and mining engineer. Became intensely involved with views of the stylised geometry of global mountain ranges (1852).

FAUJAS DE SAINT FOND, Barthélemy (1741–1819). French natural philosopher and early student of volcanic activity, ancient and modern (1778).

FISHER, Osmond (1817–1914). English cleric, geophysicist and geomorphologist. Paper on the evolution of a cliff profile (1866) remains a classic.

FORBES, James D. (1808–1869). Scottish physicist and early glaciologist.

FORSTER, Johann Reinhold (1729–1798). German naturalist, accompanied Cook on his second voyage.

GEIKIE, Archibald (1835–1924). Scottish geologist, geomorphologist and historian of geology. Christened Lyell the 'High Priest of Uniformitarianism'.

GILBERT, Grove Karl (1843–1918). Great American geologist and geomorphologist. Notable for his ability to combine the explanatory methodology of both experimental and historical science although the latter aspect has generally been downplayed by late twentieth-century workers.

GOUDIE, Andrew S. (1945–). British geomorphologist. Professor, University of Oxford. Emphasis on human impact on the Earth's surface (1981 *et seq.*) has proved extremely influential.

GOULD, Stephen J. (1941–2002). American zoologist and historian of science.

GREENWOOD, George (1799–1875). British fluvialist whose 1857 treatise represented a bitter attack on Lyell's marine interpretation of scenery.

HAYDEN, Ferdinand, V. (1829–87). Early American geologist and geomorphologist, responsible for major surveys of the Colorado Rockies and work on the balance between uplift and denudation.

HJULSTRÖM, Filip (1909–1982). Major Swedish fluvial geomorphologist, Professor, University of Uppsala. His work on the River Fyris (1935) led to the development of a classic statement of the relationship between particle size, velocity, erosion, transportation and deposition in rivers, known as the Hjulström curve.

HOLMES, Arthur (1890–1965). Brilliant British geologist and geophysicist, Professor, University of Edinburgh. Profound supporter both of radioactive dating developments and of concepts of continental drift and plate tectonics. The second edition (1965) of his textbook remains a classic.

HORTON, Robert E. (1875–1945). American engineer. Fascination with both history of geomorphology and processes of rainfall, infiltration and runoff led to great posthumous paper (1945) which acted as the springboard for late twentieth-century reductionist fluvial geomorphology.

HUMBOLDT, Alexander von (1769–1859). Extraordinarily influential German naturalist and traveller. His travels in South America (with A. Bonpland: 1799–1804) focused interest on the regularities in orientation of mountain ranges.

HUTTON, James (1726–1797). Scottish natural philosopher, Deist and originator (1785: published 1788) of the modern vision of an enormous extent of geological time. His Plutonic view of continuous renewal of land masses by the uplift of materials formed or transformed by subterranean heat was directly opposed to Werner's Neptunian concept.

HUXLEY, Thomas H. (1825–1895). English biologist and staunch upholder of Darwinian evolution. His textbook (1877) was the first to emphasize the dominant role of fluvial processes.

JAMESON, Robert (1774–1854). Scottish mineralogist. Professor, University of Edinburgh, where Darwin attended his lectures. Pupil of Werner, convinced Neptunist and scathing opponent of Hutton and Playfair.

JAMIESON, Thomas F. (1829–1913). Scottish geologist and geomorphologist. Important work on application of isostasy to glacial phenomena; correctly ascribed the Parallel Roads of Glen Roy to the role of an ice-dammed lake (1863).

JEFFERSON, Thomas (1743–1826). Polymath and third President of the USA.

JOHNSON, Douglas W. (1878–1944). American geologist and coastal geomorphologist. Pupil of Davis, whose collected essays he edited (1909). Professor, University of Columbia. Teacher of Strahler.

JUDD, John W. (1840–1916). British geologist. Editor of Darwin's geological works (1890).

JUKES, Joseph B. (1811–1869). English geologist: his work on the rivers of southwest Ireland (1862) and persistent correspondence persuaded Darwin that the Wealden Denudation and the valleys of the Blue Mountains were of fluvial rather than marine origin.

KELVIN, William Thomson, Lord (1824–1907). Major British physicist, whose opposition to Darwinian evolution led to extensive geophysical research in order to disprove the length of time claimed necessary by Darwin.

KING, Lester C. (1907–1989). South African geologist and geomorphologist. Pupil of Cotton. Professor, University of Natal. Developed a theory of back-wearing of escarpments as a counter to Davisian ideas.

KIRKBY, Michael J. (1937–). British quantitative geomorphologist. Pupil of Chorley. Professor, University of Leeds and founding editor *Earth Surface Processes and Landforms* (*né Earth Surface Processes*). Has developed a very influential modelling approach based on the continuity equation.

KIRWAN, Richard (1733–1812). Irish cleric and mineralogist. Convinced Neptunist and supporter of Biblical chronology, violently opposed to Huttonian views.

KUHN, Thomas S. (1922–1996). American physicist turned historian and philosopher of science. Professor, University of California, Berkeley; Princeton University; and Michigan Institute of Technology. Introduced (1962) the notion of 'revolutions' – sudden shifts in thinking in science – as opposed to the previous view of incremental progress.

LEOPOLD, Luna B. (1915–). American fluvial geomorphologist. U.S. Geological Survey and Professor, University of California, Berkeley. The 1964 work co-authored with Wolman and Miller was extraordinarily influential in introducing hydraulic engineering concepts to geomorphologists.

LOVELOCK, James E. (1919–). English independent scientist. His Gaia concept (1979) of the Earth as a homeostatic entity may well come to be seen to reflect the major role of bacteria and other microorganisms in Earth history.

LYELL, Charles (1797–1875). Scottish lawyer turned geologist. Briefly Professor at King's College London. His *Principles* (1830–33) were revolutionary in their insistence on scientific principles and observations. Notoriously cautious, he never fully accepted the role of rivers or land ice in Earth sculpture and was distinctly lukewarm to Darwin's evolutionary theory.

MACCULLOUGH, John (1773–1835). British geologist who seems simply to have accepted the whole range of Hutton and Playfair's ideas as common sense.

MACKIN, J. Hoover (1905–1968). American geomorphologist, notoriously sceptical of some of the quantitative methods employed, notably by Leopold.

MANDELBROT, Benoît (1924–). Polish mathematician who largely popularized the concept of fractals.

MAYR, Ernst (1904–2005). German-born biologist and historian of biology, who became the doyen of American evolutionists. Professor, Harvard University. Maintained firmly the need for distinct explanatory modes for unravelling historical phenomena.

MELTON, Mark A. (1930–). American fluvial geomorphologist and maverick. Student of Strahler. Work in 1950s and 1960s involved extensive field investigations and novel mathematical treatments which remained largely uncomprehended for 40 years or so. Retired from academic work in the late 1960s.

MILANKOVITCH, Milutin (1879–1958). Serbian physicist, who extended nineteenth century views about three interlocking astronomical cycles of solar radiation receipt into a chronology which has been fitted to Earth's temperature fluctuations during the Pleistocene.

MILLER, John P. (1923–1961). American geomorphologist, University of Harvard, who died of bubonic plague contracted during fieldwork in the USA Southwest. Co-author of the influential 1964 textbook with Leopold and Wolman.

MORISAWA, Marie E. (1918–1994). American fluvial geomorphologist. Student of Strahler. Professor, State University of New York. Co-founder, 1970, of the annual Binghamton Seminar series in geomorphology.

NEEDHAM, Joseph N. (1900–1995). British biochemist and sinologist. He initiated the mammoth *History of Science and Civilization in China* which introduced Chinese achievements in hydraulic engineering to the Western audience.

ORESKES, Naomi (1958–). American historian of Earth science. Professor, University of California, San Diego. Explores (1999) the impact of the overreliance on Chamberlin's Method of Multiple Working Hypotheses by American scientists facing Wegener's Continental Drift theory.

PALLAS, Pyotr S. (1741–1811). German natural scientist and geographer. Extensive travels in Russia produced major theories on mountains and their strata (1777–8).

PENCK, F.C. Albrecht (1858–1945). German geographer and geomorphologist. Professor, Universities of Vienna and Berlin. Responsible (with Eduard Brückner, 1901–9) for developing the classic four division model of the Ice Age: Gunz, Mindel, Riss and Würm.

PENCK, Walther (1888–1923). German geomorphologist, son of Albrecht. Developed a model of landscape evolution dominated by balance between uplift and denudation, in opposition to the Davisian Cycle.

PLAYFAIR, John (1747–1819). Scottish mathematician and natural philosopher. Professor, University of Edinburgh. Clear and elegant reformulation of Hutton's ideas (1802) represents first modern statement of the principles of fluvial geomorphology.

POPPER, Karl R. (1902–1994). Austro-British philosopher of science. Professor, London School of Economics. Originally a member of the Vienna Circle of logical positivists, Popper subsequently developed a view which placed falsifiability at the centre of scientific explanation and is closely akin to Chamberlin's ideas.

POWELL, John W. (1834–1902). American explorer, geologist and geomorphologist. Made the first European traverse of the Grand Canyon; developed key concepts such as base level.

RAMSAY, Andrew C. (1814–1891). British geologist and geomorphologist. Satisfactorily explained both the glacial origin of many deep lakes (1862) and the origin of the Wealden Denudation.

RAPP, Anders (1927–1998). Swedish geomorphologist. Professor, University of Lund. His detailed and extensive study of slope processes in Lapland (1960, 1985) remains a classic and major contribution.

RUSE, Michael (1940–). Canadian philosopher of biological science. Professor, Florida State University. His 1999 study of approaches to evolution contrasts the 'falsification' approach of Chamberlin or Popper to 'normal' science studies, with that of Kuhnian 'revolutionary' acceptance or rejection of paradigm shifts.

RUTHERFORD, Ernest, Baron (1871–1937). New Zealand physicist. Professor at McGill and Cambridge Universities. Put forward (1904) the concept that radioactive decay permitted the extension of Earth's age beyond the period deduced by Kelvin.

SAUSSURE, Horace B. de (1940–1999). Swiss glaciologist and geomorphologist. His monumental description of Alpine scenery (1779–96) made clear the shortcomings of simple fluvial descriptions, such as Bourguet's or Hutton's.

SCHEIDEGGER, Adrian (1925–). Austrian geomorphologist. Professor, Technical University of Vienna. His 1961 text represented the first wholesale modern physical and mathematical treatment of landforms.

SCHUMM, Stanley A. (1927–). Major American fluvial geomorphologist. U.S. Geological Survey, then Professor, University of Colorado, Boulder. Coherent view of fluvial processes given by his 1977 text, incorporates both historical and experimental approaches. Concept of geomorphic thresholds and complex response especially important.

SCROPE, George J.P. (1797–1876). English geologist. Early conviction of the efficacy of fluvial erosion as demonstrated among the volcanoes of France (1827) met little acceptance.

SELBY, Michael J. (1936–). English–New Zealand geomorphologist, Professor, University of Waikato. Major studies of slope forms (1993) especially in Antarctica; developed the concept of rock-mass strength equilibrium slopes.

SHREVE, Ronald L. (1930–). American geomorphologist. Professor, University of California, Los Angeles. His concept of topologically distinct channel networks (1966) advanced Hortonian ideas into mathematically interesting directions.

SIMPSON, George G. (1902–1984). American palaeontologist. His discussion (1963) of the differences between explanatory modes in experimental and historical sciences remains classic.

STODDART, David R. (1938–). English coastal geomorphologist and historian of geomorphology. Professor, University of California, Berkeley. Enthusiastic upholder of the historical, natural philosophy tradition.

STRAHLER, Arthur N. (1918–2002). American geomorphologist. Pupil of Johnson, Professor, University of Columbia. Developed the first school of quantitative geomorphology, with significant publications in 1950 and 1952, the latter effectively advocating physical over historical approaches to landform studies.

TAYLOR, Thomas J. (1810–1861). English mining engineer, whose 1851 hydraulics text foreshadows much of Leopold's later work.

TRICART, Jean L.F. (1920–2003). Highly influential French Marxist geomorphologist. Professor, University of Strasbourg. Long-term editor, *Révue de géomorphologie dynamique*.

TYNDALL, John (1820–1893). Irish glaciologist.

USSHER, James (1581–1656). Biblical scholar and Archbishop of Armagh. Constructed a meticulous chronology based on Biblical and other sources, which placed the date of the Creation in 4004 BC.

WEGENER, Alfred L. (1880–1930). German meteorologist, Arctic explorer and proponent of the theory of Continental Drift (1915, 1924), the forerunner of modern plate tectonics.

WERNER, Abraham G. (1749–1817). German mineralogist and natural philosopher. Professor School of Mines, Saxony. Developed what became the very influential Neptunist Theory whereby all rocks except modern lavas were seen as precipitated by the Noachian floodwaters.

WOLMAN, M. Gordon 'Reds' (1924–). American fluvial geomorphologist. Professor, Johns Hopkins University. Co-author of influential 1964 text with Leopold and Miller.

WOOLDRIDGE, Sidney W. (1900–1961). English geomorphologist and major supporter of Davisian interpretations of southern English scenery. Professor, King's College, London.

Appendix II

Antecedent stream: defined by Powell. One which remains fixed in its course as uplift occurs across its valley. Major Himalayan river courses are antecedent.

Base level: defined by Powell. The lower limit of subaerial erosion by running water. May be local (for example, a lake) or universal (the sea). Glaciers and mass movements may erode below base level.

Catastrophism: an outdated version of Earth history in which major landforming events are ascribed to Divine intervention (for example, Noah's Flood).

Complex response: concept promoted by Schumm. Key example was experimental evidence that a single fall in *base level* led: first, to a wave of erosion, incision and formation of one pair of terraces; second, to the deposition of eroded material to partly infill the eroded valley floors; thirdly, to partial re-excavation of the fill as the channel profile finally adjusted, creating a second, lower pair of terraces. Thus instead of one set of paired terraces, two sets resulted, from one change in relative relief and stream power.

Continental drift: concept introduced by Wegener not merely to explain the mirror-image 'fit' of land masses on either side of the Atlantic, but also to account for the distribution of a range of topographic, geological and palaeo features. Incorrectly originally assumed to be occurring at metres/year. Formed the background to *plate tectonics* (millimetres/year).

Deism: a belief which became common in eighteenth-century Europe, that rejected the Bible as the source of information concerning the existence and actions of a Supreme Being. Instead, evidence for the working of a Divine Providence was thought to be provided by the natural world itself.

Drainage density: Concept introduced by Horton for the quantitative expression of the degree of fluvial dissection of drainage basins. Expressed as the area of the basin divided by the length of stream channels, or valleys. Its relative scale-invariance may represent a *fractal* condition.

Endogenetic/endogenic: those processes operating below the surface of a planet, governing uplift and subsidence. On Earth, tectonism and volcanism.

Enlightenment, The: term given to the rise of scientific, rational approaches to the study of Man and Nature in Europe in the mid-eighteenth century. Key protagonists in France commenced with Voltaire and included the *Encyclopédistes* such as

Diderot. The notable flowering in Edinburgh created a major group which in-cluded Hutton and Playfair as well as Sir James Hall, the chemist.

Exogenetic/exogenic: forces which operate from above to alter a planet's surface. Include meteorite impacts, wind, weathering, precipitation, rivers, glaciers and the sea.

Experimental science (or physical science): classic approach to exploration of cause and effect by means of controlled experimentation. Here all variables but one will be held constant ('fixed boundary conditions') and the response to differences in nature of the sole remaining variable are used to construct mathematical rela-tionships and laws (for example, the Gas Laws).

Eustasy: sea-level change on a global scale. Caused principally by changes in the volume of the ocean basins, or – and more commonly during the *Quaternary period* – by changes in the volume of sea water, as more or less is locked up in land-based ice sheets and glaciers.

Fractal or fractal dimension: concept developed by Mandelbrot which expresses the way in which curved lines fill space, measured as a dimension between 1.0 and 2.0. Fractals permit the mathematical characterization of the apparently irregular horizontal and vertical distributions of elements of natural topography. The original study of the British coast, measured at a variety of scales, showed a completely scale-free, apparently indefinitely-repeated fractal dimension.

General Systems Theory: concept developed by Bertalanffy in the hope of unifying the physical and biological sciences and imported into geomorphology by Chorley. The focus is on the nested hierarchies of phenomena and of the flows of energy (in many forms) and matter between them.

Geographical Cycle: W.M. Davis's brilliant descriptive device which brought the development of landscapes over time into a sequence of youthful, mature and old-age forms, culminating in the virtually flat *peneplain*. Ostensibly Darwinian in its inspiration, the cycle is more accurately a historical sequence.

Geological Cycle: Hutton's recognition of the apparently incessant transfer of material from landmass to ocean and, after conversion to rock, uplift to create new landmasses.

Grade: a theoretical condition of balance between processes of subaerial erosion and deposition, proposed by Gilbert. Its link to the physical gradient of slopes and streams created substantial practical confusion.

Historical (or observational) science: the study of cases in which the specific and unique contingencies of time and place have determined the sequential develop-ment of phenomena. Although mathematical modelling can assist in creating sce-narios which may shed light on the likelihood or otherwise of particular sequences, the nature of scientific proof in explanations in historical science is more akin to that in a courtroom, and rests on the balance of probabilities. Darwinian evolution is a major theory of historical science whose proof (or support) rests on quite different, but equally scientific, bases to the Gas Laws.

Hysteresis: the time- and space-specific phenomenon which links different levels of response to similar levels of applied force. For example, two river floods of the same discharge at the same location will move different amounts of sediment if they occur close together or far apart, as the sediment supply may be very different in each case.

Isostasy: the phenomenon displayed by the rather rigid outer layer of the Earth's crust, floating upon the underlying mantle rocks. There is a hydrostatic equilibrium between crust and mantle which is perturbed by the differential loading and unloading of the crust both by the erosion and deposition of sediments and by the growth and decay of major ice sheets.

Method of Multiple Working Hypotheses: Chamberlin's formulation of an older method for evaluating the viability of competing explanations. This is especially relevant to studies in *historical science*, where critical, direct experiments are generally impossible. It requires the development of a set of potential explanations which are then scrupulously tested against the evidence available. It may be possible to specify the nature of critical evidence, if no complete elimination of competing ideas is possible. It is characteristic of periods of *'normal' science* and fits Popper's idea of falsifiability as the key to developing scientific explanation.

Milankovitch (or Croll–Milankovitch) mechanism: the hypothesis that changes in the Earth's astronomical relationship to the Sun create cyclic differences in the receipt of solar radiation which, in turn, drive the sequence of glacial and interglacial periods within an Ice Age. There are three sets of variations: in the eccentricity of the orbit (with a period of 96,000 years); in the *precession of the equinoxes* (a period of 21,000 years); and the *obliquity of the ecliptic* (40,000 years). The mutual timing of these cycles can reinforce or minimize changes in radiation receipt. The calculation of these effects is complex and there is rather limited long-term evidence that they are truly 'the pacemaker of the Ice Age', since the maximum differences in temperature produced may be only 1° or 2°. Argument continues both over the efficacy of these processes and how – if at all – they link to shorter, 'sub-Milankovitch' variations.

Mosaic chronology: as described in the Biblical Book of Genesis, which was thought to represent the account God imparted to the prophet Moses. Gave a period of six days for the Creation of the Earth and its inhabitants. Ussher dated the Creation itself to 4004 BC using the Bible as his primary source.

Noachian deluge (or Noah's Flood): the period of apparently global and wholesale inundation described in Genesis. It was widely held that this had substantially modified the topography of the Creation. Where the water went afterwards was a puzzle which required ingenious solutions.

Neo-catastrophism: a late twentieth century attempt to focus attention on the geomorphological power of events which are relatively rare on the timescale of human memory (for example, the Indian Ocean tsunami of 26 December 2004). Whilst a sensible attempt to redirect attention from the very short-term, small-scale phenomena which dominated much Anglo-American work in the 1960s–1980s, the term is misleading, as it has no connection with the far earlier *Catastrophist* concept.

Neptunism: the eighteenth century concept associated particularly with Werner which considered all rocks apart from modern lavas had been precipitated from the waters either of the Creation or Noah's Flood.

Non-linear dynamics: descriptive of natural systems that do not settle to a simple, equilibrium state which can be readily expressed mathematically. Although most natural systems are non-linear, many can be approximated by linear mathematical formulae. However, many – and this includes meteorological phenomena – show what is termed chaotic behaviour with multiple possible relationships between the level of a controlling variable and the behaviour of the phenomenon (cf. *hysteresis*).

Normal (or puzzle-solving) science: Kuhn's description of conditions when a scientific field possesses a well-established *paradigm* and investigations proceed using clearly understood criteria for acceptance and rejection of evidence.

Obliquity (of the ecliptic): the angle between the plane of the Earth's orbit and the plane of its rotational equator. It varies with a time period of about 40,000 years.

Paradigm: Kuhn's term for the guiding principles of scientific belief at any period of *normal science*. Many shades of definition have been noted, but in essence a paradigm encompasses a set of high-level concepts which are not challenged by the everyday operations of scientists. Two paradigms generally cannot comfortably co-exist in one science – as during periods of *scientific revolutions* – since results obtained under one are incommensurable with the tenets of the other.

Peneplain: the presumed endpoint of the *Geographical Cycle*. A gently-sloping surface, with limited areas of high relief, termed monadnocks. Evidence of widespread *unconformities* support the possibility of the production of such flat surfaces by subaerial processes, but there was endless debate in the mid-twentieth century about which processes would produce what type of surface.

Plate tectonics: development of Wegener's theory of *Continental Drift*, largely formulated and generally accepted during the late 1950s–1960s. There appear to be a number of rigid crustal plates which move over the Earth's mantle rocks probably on convection currents. Where the plates meet, there are major tectonic phenomena, including earthquakes and volcanic activity. The Indian Ocean *tsunami* of 26 December 2004 was caused by a massive rupture along part of the boundary between the Indian and Eurasian plates, where the former is being driven under ('subducting beneath') the latter.

Plutonism: the view developed by Hutton, in contrast to Neptunist ideas, which held that most rocks were formed by the action of subterranean heat and pressure, then raised to the surface by *endogenetic* forces.

Quantitative revolution: a period centred from the 1950s to 1970s in Anglo-American geography, in particular, when the historical, narrative and largely qualitative approaches to both natural and human phenomena were (temporarily) overturned in favour of mathematical analysis and modelling. In geomorphology, Strahler and Chorley were especially important. The approach has largely given way to more profound reliance upon geophysical techniques (see Church, 2005).

Quaternary: the latest geologic period, including the Pleistocene (or Ice Age) and the Holocene or Recent, after the retreat of the last major ice sheets. There is continuing debate about the precise boundaries of both, but the Pleistocene began about 2.7 million years ago and the Holocene about 12,000. There is no way of telling whether the Ice Age is over or whether the Holocene is merely an interglacial warm period.

Reductionism: or the principle of structural explanation. Considers that the properties of phenomena should be reduced to a limited number of underlying elementary units or laws, usually of physics or chemistry. Phenomena with unique histories are fitted poorly by reductionist explanations.

Sand-pile theory: a concept of Bak and co-workers to describe the non-linear relation between event and outcome. If a single sand grain is dropped on a stable pile of sand, anything may happen, between the movement of one grain and a wholesale collapse. The distribution of results matches the distribution of events such as earthquakes, but there is no way of predicting the size of the result in any one case.

Scientific revolution: Kuhn's concept of sudden changes in dominant *paradigms*: an example would be the substitution of *plate tectonics* for the view of an immobile Earth. Kuhn considers that the precursor of a revolution is often the accumulation of anomalies, leading to dissatisfaction with the continuing explanatory power of the existing paradigm. It is, however, not possible to predict how or when a revolutionary new paradigm may be proposed and gain acceptance.

Self-organizing criticality: the description of phenomena which appear to exhibit chaotic behaviour, but revert from time to time to 'central' states. Stølum's description of the behaviour of meandering streams is an excellent example. There is no need to invoke external forces, to account for dramatic switches in configuration.

Superimposition: (of drainage) The case where a valley network is established on an upper surface and is then lowered by erosion onto an underlying set of beds, where the network appears completely alien to the geologic structure of the lower surface. A classic example is the English *Weald*, where drainage developed on a gently sloping upper surface has been 'let down' onto the underlying dome of the Wealden rocks. One result is that small streams flow through the ring of high ground represented by the North and South Downs.

Threshold: a discontinuity in the development of a phenomenon. In many cases in earth science, systems are driven across thresholds by changes in external controls: for example, increased precipitation intensity may result in new sets of gullies as the infiltration thresholds of hillsides are exceeded. Geomorphic thresholds, by contrast, Schumm considers an inherent part of the operation of natural processes: the cutting-off of a meander, to form an oxbow lake is an intrinsic feature of the meandering process.

Tsunami: a large and violent ocean wave or series of waves triggered by an underwater disturbance, such as earthquake, volcanic eruption or landslide. Tsunamis are virtually undetectable in the open ocean and can travel at speeds of hundreds of kilometres per hour. The borders of the Pacific Ocean are especially prone to tsunamis, but they may occur anywhere and are particularly destructive when funnelled into shallow bays.

Unconformity (geological): a clear discontinuity in the attitude of successive layers of rock, which indicates an episode of erosion or interruption to sedimentation. Hutton's Jedburgh example remains classic.

Uniformitarianism: a much misunderstood tenet of all natural sciences, which is particularly associated with the views of Lyell. In essence, it is no more than a scientific precept which requires the exhaustion of all possible, known causes before having recourse to new or improbable explanations for phenomena. Best summed up by Sherlock Holmes' dictum: 'When you have eliminated the impossible, whatever is left, however improbable, must be the truth.' It has nothing whatsoever to do with the 'uniformity' or 'gradualness' of the operation of processes. (See Shea's paper.)

Vade mecum: (Latin phrase: 'to go with me') a guide, especially a guide book.

Wealden Denudation: the Weald of Southeast England is an eroded dome, surrounded by inward-facing Chalk escarpments. Small streams such as the River Darenth, flow out from the centre of the Weald, through improbably large valleys in the Chalk. This topography was demonstrated by Ramsay to be produced by the *superimposition* of the drainage from an original Chalk covering. Darwin's earlier estimate of the time needed for the denudation by marine action led to Kelvin's attempts to prove, by physical principles, that the Earth's history was too short for the length of time Darwin proposed.

Whig (or Whiggish) history: a now-discredited reading of historical events such that they are shown to lead, inexorably and inevitably, to the present state of things. The latter is also generally viewed with complacency as the best of all possible outcomes.

References

Adams, D., 1980. *The Hitchhiker's Guide to the Galaxy*. London: Pan Books.

Agassiz, J.L.R., 1840a. *Etudes sur les glaciers*. Neuchâtel: privately published.

Agassiz, J.L.R., 1840b. On glaciers, and the evidence of their having once existed in Scotland, Ireland, and England. *Proceedings, Geological Society of London*, 3, 327–32.

Ahnert, F., 1987. Approaches to dynamic equilibrium in theoretical simulations of slope development. *Earth Surface Processes and Landforms*, 12, 3–15.

Anderson, M.G. & Calver, A., 1977. On the persistence of landscape features formed by a large flood. *Transactions, Institute of British Geographers* (NS), 2, 243–54.

Bagrow, L., 1985. *History of Cartography* (2nd edn, revised and enlarged, R.A. Skelton). Chicago: Precedent Publishing Inc.

Bak, P., 1996. *How Nature Works: the Science of Self-organised Criticality*. New York: Copernicus Press.

Baker, V.R. (ed.), 1981. *Catastrophic Flooding*. Stroudsburg: Dowden, Hutchison and Ross.

Baker, V.R., 1993. Extraterrestrial geomorphology: science and philosophy of Earthlike planetary landscapes. *Geomorphology*, 7, 9–35.

Baker, V.R., Kochel, R.C. & Patton, P.C. (eds), 1988. *Flood Geomorphology*. New York: Wiley Interscience.

Bandfield, J.L., Glotch, T.D. & Christensen, P.R., 2003. Spectroscopic identification of carbonate minerals in the Martian dust. *Science*, 301, 1084–7.

Barry, R.G., 1967. Models in meteorology and climatology. In: R.J. Chorley and P. Haggett (eds), *Models in Geography*. London: Methuen, 97–144.

Barry, R.G., 1997. Palaeoclimatology, climate system processes and the geomorphic record. In: D.R. Stoddart (ed.), *Process and Form in Geomorphology*. London: Routledge, 187–214.

Basset, C., 1815. (Translated and edited.) *Explication de Playfair sur la théorie de la Terre par Hutton*. Paris: Bossange et Masson.

Baxter, S., 2003. *Revolutions in the Earth*. London: Weidenfeld & Nicolson.

Beckinsale, R.P. & Chorley, R.J., 1991. *The History of the Study of Landforms, III: Historical and Regional Geomorphology 1890–1950*. London: Routledge.

Bennett, R.J. & Chorley, R.J., 1978. *Environmental Systems: Philosophy, Analysis and Control*. London: Methuen.

Bertalanffy, L. von, 1950. The theory of open systems in physics and biology. *Science*, 111, 23–29.

Biram, J. (translator and editor), 1966. *A. Wegener: the Origin of Continents and Oceans (4th edn), 1929*. New York: Dover Publications.

Bishop, P. & Cowell, P., 1997. Lithological and drainage network determinants of the character of drowned, embayed coastlines. *Journal of Geology*, 105, 685–700.

Blyth, K. & Rodda, J.C., 1973. A stream length study. *Water Resources Research*, 9, 1454–1461.

Boardman, J., 2003. Soil erosion and flooding on the eastern South Downs, Southern England, 1976–2001. *Transactions, Institute of British Geographers* (NS), 28, 176–96.

Bonney, T.G., 1896. *The Story of our Planet*. London: Cassell.

Bourguet, L., 1729. *Lettres philosophiques sur la formation des sels et des cristaux . . . Mémoire sur la théorie de la Terre*. Amsterdam: François l'Honoré.

Bourguet, L., 1742. *Traité des petrifications*. Paris: Briasson.

Bowen, D.Q., 2004. Ice ages (interglacials, interstadials and stadials). In: A.S. Goudie (ed.), *Encyclopedia of Geomorphology* (2 vols). London: Routledge, 549–54.

Breislak, S., 1798. *Topographia fisica della Campania di Roma*. Firenze.

Breislak, S., 1811. *Introduzione alla geologia*. Milano: Stamperia Reale (2 volumes).

Breislak, S., 1812. *Introduction à la géologie, ou à l'histoire Naturelle de la terre*. Paris: J. Klosterman fils.

Breislak, S., 1818. (Translated, P.J.L. Campmas.) *Institutions géologiques*. Milan: Imprimérie impériale et royale (3 volumes + Atlas).

Broecker, W.S., 2000. Was a change in the thermohaline circulation responsible for the Little Ice Age? *Proceedings, National Academy of Sciences, USA*, **97**(4), 1339–1352.

Broecker, W.S., 2003. Does the trigger for abrupt climate change reside in the ocean or in the atmosphere? *Science*, **300**, 519–22.

Broecker, W.S., 2004. Future global warming scenarios, *Science*, **304**, 388.

Browne, J., 1995. *Charles Darwin: Voyaging*. London: Jonathan Cape.

Browne, J., 2002. *Charles Darwin: the Power of Place*. London: Jonathan Cape.

Bryson, B., 2003. *A Short History of Nearly Everything*. London: Doubleday.

Buckland, W., 1823. *Reliquiae Diluvianae or, Observations on the Organic Remains Contained in Caves, Fissures and Diluvial Gravel, and on other Geological Phenomena, Attesting the Action of an Universal Deluge*. London: John Murray (2nd edn, 1824).

Buckland, W., 1840–1. On the evidences of glaciers in Scotland and the north of England. *Proceedings, Geological Society*, **3** (part 1), 332–7; (part 2) 345–8.

Buffon, G.L.L., 1749. *Histoire Naturelle, générale et particulière, avec la description du Cabinet du Roi*, Vol. I. Paris: L'imprimérie royale.

Buffon, G.L.L., 1753. *Histoire Naturelle, générale et particulière, avec la description du Cabinet du Roi*, Vol. IV. Paris: L'imprimérie royale.

Buffon, G.L.L., 1775. *Histoire Naturelle, générale et particulière, servant de suite de la Théorie de la terre, parties expérimentale et hypothétique*. Paris: L'imprimérie Royale.

Buffon, G.L.L., 1778. *Histoire Naturelle, générale et particulière supplément: Des Époques de la Nature*. Paris: Imprimérie Royale.

Burchfield, J.D., 1990. *Lord Kelvin and the Age of the Earth* (2nd edn). Chicago: University of Chicago Press.

Burkhardt, F. & Smith, S. (eds), 1990. *The Darwin Correspondence*, Vol. 6, 1856–7. Cambridge: Cambridge University Press.

Burkhardt, F., Browne, J., Porter, D.M. & Richmond, M. (eds), 1993. *The Darwin Correspondence*, Vol. 8, 1860. Cambridge: Cambridge University Press.

Burkhardt, F., Browne, J., Porter, D.M. & Richmond, M. (eds), 1994. *The Darwin Correspondence, 1861*. Vol. 9. Cambridge: Cambridge University Press.

Burkhardt, F., Harvey, J., Porter, D.M. & Topham, R. (eds), 1997. *The Darwin Correspondence*, Vol. 10, 1862. Cambridge: Cambridge University Press.

Burkhardt, F., Porter, D.M., Dean, S.A., Topham, R. & Wilmot, S. (eds), 1999. *The Darwin Correspondence*, Vol. 11, 1863. Cambridge: Cambridge University Press.

Campbell, N.A., 1996. *Biology* (4th edn). Menlo Park: The Benjamin/Cummings Publishing Co. Inc.

Carozzi, A.V. (translator and editor), 1967. *Studies on Glaciers: Preceded by the Discourse of Neuchâtel by Louis Agassiz*. New York: Hafner.

Carter, L.J., 1980. The Leopolds: a family of naturalists. *Science*, **207**, 1051–5.

Chamberlin, T.C., 1890. The method of multiple working hypotheses. *Science*, **15**, 92–96.

Charpentier, J. de, 1841. *Essai sur les glaciers et le terrain érratique du Bassin du Rhône*. Lausanne.

Chorley, R.J., 1962. Geomorphology and General Systems Theory. *U.S. Geological Survey Professional Paper*, **500B**.

Chorley, R.J. & Haggett, P. (eds), 1965. *Frontiers in Geographical Teaching*. London: Methuen.

Chorley, R.J. & Haggett, P. (eds), 1967. *Models in Geography*. London: Methuen.

Chorley, R.J. & Kennedy, B.A., 1971. *Physical Geography: a Systems Approach*. London: Prentice Hall.

Chorley, R.J., Dunn, A.J. & Beckinsale, R.P., 1964. *The History of the Study of Landforms*. Vol. I. *Geomorphology before Davis*. London: Methuen.

Chorley, R.J., Beckinsale, R.P. & Dunn, A.J., 1973. *The History of the Study of Landforms*. Vol. II. *The Geomorphology of William Morris Davis*. London: Methuen.

Church, M.A., 2005. Continental drift. *Earth Surface Processes and Landforms*, **30**, 129–130.

Church, M.A. & Mark, D.A., 1980. On size and scale in geomorphology: *Progress in Physical Geography*, **4**, 342–390.

Clarke, G., Leverington, D., Teller, J. & Dyke, A. 2003. Superlakes, mega floods, and abrupt climate change. *Science*, **301**, 922–3.

Coates, D.R. & Vitek, J.D. (eds), 1980. *Thresholds in Geomorphology*. London: George Allen and Unwin.

Cohen, I.B., 1985. *Revolution in Science*. Cambridge, Mass: Belknap Press.

Conan Doyle, A., 1890. *The Sign of Four*. London: John Murray.

Cotton, C.A., 1945. *Geomorphology: an Introduction to the Study of Landforms* (4th edn). Christchurch: Whitcombe and Tombs.

Crutzen, P.J. & Stoermer, E.F., 2000. The 'Anthropocene'. *Global Change Newsletter*, **41**, 12–3.

Cunningham, F., 1990. *James David Forbes: Pioneer Scottish Glaciologist*. Edinburgh: Scottish Academic Press.

Cuvier, G., 1810. *Rapport historique sur le progrès des Sciences Naturelles depuis 1789, et sur leur état actuel*. Paris: Imprimérie impèriale.

Cuvier, G., 1825. *Discours sur les révolutions de la surface du Globe et sur les changemens qu'elles ont produit dans le règne animal* (3rd edn). Paris: Dufour et Ocagne.

Dadson, S.J., Hovius, N., Chen, H., Dade, W.S., Hsieh, M-L., Willett, S.D., Hu, J-C., Horng, M-J, Chen, M-C., Stark, C.P., Lagun, D. & Lin, J-C., 2003. Links between erosion, runoff variability and seismicity in the Taiwan orogen. *Nature*, **426**, 648–51.

Dana, J.D., 1863. *Manual of Geology*. Philadelphia: Bliss and Co.

Dana, J.D., 1869. *Manual of Geology* (2nd edn). Philadelphia: Bliss & Co.

Darwin, C.R., 1839a. *Journal of researches into the geology and natural history of the various countries visited by H.M.S. 'Beagle'* London: (Reprinted variously as *A Naturalist's Voyage* and *The Voyage of the 'Beagle'*.)

Darwin, C.R., 1839b. Observations on the parallel roads of Glen Roy. *Philosophical Transactions Royal Society London*, **Part I**, 39–81.

Darwin, C.R., 1842. *The Structure and Distribution of Coral Reefs. Part I of The Geology of the Voyage of the 'Beagle'*. London: Smith, Elder & Co.

Darwin, C.R., 1844. *Geological Observations on the Volcanic Islands visted during the Voyage of H.M.S. 'Beagle'*. London: Smith and Elder.

Darwin, C.R., 1846. *Geological Observations on South America. Part 3 of The Geology of the Voyage of the 'Beagle'*. London: Smith and Elder.

Darwin, C.R., 1859. *On the Origin of Species*. London: John Murray.

Darwin, C.R., 1860a. *On the Origin of Species* (2nd edn). London: John Murray.

Darwin, C.R., 1860b. *Journal of Researches*. Reprint of 1845, 2nd edn with Postscript. London: John Murray.

Darwin, C.R., 1869. *On the Origin of Species* (5th edn). London: John Murray.

Darwin, C.R., 1881. *The formation of vegetable mould by the action of earthworms with observations on their habits*. London: John Murray.

Davies, G.L., 1968. *The earth in decay: a history of British Geomorphology 1578–1878*. London: Macdonald & Co.

Davis, W.M., 1888. Wasp stings. *Science*, **11**, 50.

Davis, W.M., 1894. *Elementary Meteorology*. Boston: Ginn & Co.

Davis, W.M., 1921. Airplane views of the Alps. *Journal of Geography*, **20**, 116–7.

Davis, W.M., 1926. The value of outrageous geological hypotheses. *Science*, **63**, 463–8. (Reprinted, 1975, in C.C. Albritton (ed.)), *Philosophy of Geohistory*. Stroudsburg: Dowden, Hutchinson and Ross).

Davis, W.M., 1938. Sheetfloods and streamfloods. *Geological Society of America, Bulletin*, **49**, 1337–416.

Dean, J.R., 1992. *James Hutton and the History of Geology*. Ithaca: Cornell University Press.

De la Noë, G.D. & de Margerie, E., 1888. *Les formes du terrain*. Paris: Imprimérie Nationale.

De Luc, J.A., 1810–11. *Geological Travels*. London: F.C. & J. Rivington (3 volumes).

Diamond, J., 2005. *Collapse: How Societies Choose to Fail or Survive*. London: Allen Lane.

Du Toit, A.L., 1937. *Our Wandering Continents*. Edinburgh: Oliver and Boyd.

Dunne, T. & Leopold, L.B., 1978. *Water in Environmental Planning*. San Francisco: W.H. Freeman.

Dury, G.H., 1964–5. Studies of underfit streams. *U.S. Geological Survey, Professional Paper*, **452A–C**.

Dutton, C.E., 1882. *Tertiary History of the Grand Canyon Region*. Washington: Monographs, U.S. Geological Survey, 2.

Elie de Beaumont, L., 1852. *Notice sur les systèmes de montagnes*, 3 vols. Paris: P. Bertrand.

Faujas de Saint Fond, B., 1778. *Recherches sur les volcans éteints du Vivarais et du Velay* . . . Grenoble: J. Cuchet.

Fenneman, N.M., 1936. Cyclic and non-cyclic aspects of erosion. *Science,* **83**, 87–94.

Feyerabend, P., 1975. *Against Method: Outline of an Anarchistic Theory of Knowledge.* London: NLB.

Fisher, O., 1866. On the disintegration of a chalk cliff. *Geological Magazine,* **3**, 354–6.

Forster, J.R., 1778. *Observations Made during a Voyage Round the World.* London: G. Robinson.

Geikie, A., 1905. *The Founders of Geology* (2nd edn). London: Macmillan & Co (1st edn, 1897).

Gilbert, G.K., 1876. The Colorado Plateau Province as a field for geological study. *American Journal of Science,* **68**, 16–24; 85–103.

Gilbert, G.K., 1877 (published 1879). *Report on the Geology of the Henry Mountains.* Washington: Government Printing Office.

Gilbert, G.K., 1885. The topographic features of lake shores. *Geological Survey, Annual Report,* 5. Washington, 69–123.

Gilbert, G.K., 1886. The inculcation of the scientific method by example. *American Journal of Science* (3rd series), **31**, 284–99.

Gilbert, G.K., 1890. Lake Bonneville. *U.S. Geological Survey, Monograph,* **1**, 23–65.

Gilbert, G.K., 1893. The Moon's face; a study of the origin of its features. *Philosophical Society of Washington, Bulletin,* **12**, 241–92.

Gilbert, G.K., 1896. The origin of hypotheses, illustrated by the discussion of a topographic problem. *Science,* **3**, 1–13.

Gilbert, G.K., 1909. The convexity of hill-tops. *Journal of Geology,* **17**, 344–50.

Gilbert, G.K., 1914. On the transportation of debris by running water. *U.S. Geological Survey Professional Paper,* **86**.

Gilbert, G.K., 1917. Hydraulic mining debris in the Sierra Nevada. *U.S. Geological Survey Professional Paper,* **105**.

Giusti, C., 2004. Géologues et géographes français face à la théorie davisienne (1896–1909): retour sur 'l'intrusion' de la géomorphologie dans la géographie. *Géomorphologie,* **3**, 241–254.

Gleick, J., 1987. *Chaos: Making a New Science.* London: Vintage.

Goetzmann, W.H. & Sloan, K., 1982. *Looking far North: the Harriman Expedition to Alaska, 1899.* Princeton: Princeton University Press.

Goudie, A.S., 1981. *The Human Impact: Man's Role in Environmental Change.* Oxford: Blackwell.

Goudie, A.S. (ed.), 2004. *Encyclopedia of Geomorphology.* London: Routledge (2 volumes).

Gould, S.J., 1977. *Ontogeny and Phylogeny.* Cambridge, Mass: The Belknap Press.

Gould, S.J., 1987. *Time's Arrow, Time's Cycle: Myth and Metaphor in the Discovery of Geological Time.* Cambridge, Mass: Harvard University Press.

Greenwood, G., 1857. *Rain and Rivers; or, Hutton and Playfair versus Lyell and All Comers.* London: Longman, Brown, Green.

Gregory, K.J., 1966. Dry valleys and the composition of the drainage net. *Journal of Hydrology,* **4**, 327–340.

Gregory, K.J. & Walling, D.E., 1973. *Drainage Basin Form and Process: a Geomorphological Approach.* London: Edward Arnold.

Gross, M., 2001. *Life on the Edge: Amazing Creatures Thriving in Extreme Environments,* Cambridge, MA: Perseus Publishing.

Hall, N., 1992. *The 'New Scientist' Guide to Chaos.* London: Penguin Books.

Hartt, C.F., 1870. *The Thayer Expedn: Scientific Results of a Journey in Brazil by Louis Agassiz and his Travelling Companions.* Boston: Fields, Osgood.

Hayden, F.V., 1862. Some remarks in regard to the period of elevation of those ranges of the Rocky Mountains . . . *American Journal of Science* (2nd series), **33**, 305–13.

Heffner, A., 1928. *Die Oberflächenformen des Festlandes* (2nd edn). Leipzig and Berlin: Teubner. (Translated into English by P. Tilley, 1972, as *The Surface Features of the Earth.* London: Macmillan.)

Hilborn, R. & Mangel, M. 1997. *The Ecological Detective: Confronting Models with Data.* Princeton: Princeton University Press.

Hjulström, F., 1935. Studies of the morphological activities of rivers as illustrated by the River Fyris. *Geological Institute of Uppsala Bulletin,* **25**, 221–527.

Holmes, A., 1965. *Principles of Physical Geology* (2nd edn). London: Nelson.

Hooke, R. le B., 1999. Spatial distribution of human geomorphic activity in the United States: comparison with rivers. *Earth Surface Processes and Landforms*, **24**, 687–92.

Hooykaas, R., 1970. *Catastrophism in Geology, its Scientific Character in Relation to Actualism and Uniformitarianism*. Amsterdam: North-Holland Publishing Co.

Horton, R.E., 1933. The role of infiltration in the hydrological cycle. *Transactions, American Geophysical Union*, **14**, 446–60.

Horton, R.E., 1945. Erosional development of streams and their drainage basins: hydrophysical approach to quantitative morphology. *Geological Society of America, Bulletin*, **56**, 275–370.

Huggett, R.J., 2003. *Fundamentals of Geomorphology*. London: Routledge.

Humboldt, A. von, 1823. *A Geognostical Essay on the Superposition of Rocks in Both Hemispheres*. London: Longman, Hurst, Rees, Orme, Brown and Green.

Humphreys, A.A. & Abbott, H.L., 1861. Report on the physics and hydraulics of the Mississippi River. *U.S. Corps of Topographical Engineers, Professional Paper*, **4**.

Hunt, C.B., 1980. G.K. Gilbert's Lake Bonneville studies. In: E.L. Yochelson (ed.). *The Scientific Ideas of G.K. Gilbert. U.S. Geological Survey Special Paper*, **183**, 45–59.

Hurst, H.E., 1950. Long-term storage capacity of reservoirs. *American Society of Civil Engineers, Transactions*, **116**, 770–808.

Hutton, J., 1788. Theory of the Earth. *Royal Society Edinburgh, Transactions*, **1**, 209–304.

Hutton, J., 1795. *Theory of the Earth, with Proofs and Illustrations*. Vols. I & II. Edinburgh: Cadell, Junior, Davies & Creech. (Vol. III, edited by A. Geikie, London, 1899.)

Huxley, T.H., 1877. *Physiography: an introduction to the study of Nature*. London: Macmillan.

Imbrie, J. & Imbrie, K.P., 1979. *Ice Ages: Solving the Mystery*. London: Macmillan.

Jameson, R., 1808. *System of Mineralogy*, Vol. III, *Elements of Geognosy*. Reprinted, in facsimile, 1976 as: *The Wernerian Theory of the Neptunian Origin of Rocks* (edited by G.W. White). New York: Hafner.

Jamieson, T.F., 1863. On the parallel roads of Glen Roy, and their place in the history of the Glacial Period. *Quarterly Journal, Geological Society, London*, **19**, 235–59.

Jannasch, H.W. & Mottl, M.J., 1985. Geomicrobiology of deep-sea hydrothermal vents. *Science*, **229**, 717–25.

Jefferson, T., 1788. *Notes on the State of Virginia*. Philadelphia: Prichard and Hall.

Johnson, D.W. (ed.), 1909. *W.M. Davis, Geographical Essays*. Boston: Ginn and Co.

Johnston, A.E., 1994. The Rothamsted Classical Experiments. In: R.A. Leigh and A.E. Johnston (eds), *Long-term Experiments in Agricultural and Ecological Sciences*. Wallingford: CAB International, 9–37.

Jones, D.K.C., 1981. *The Geomorphology of the British Isles: South-east and Southern England*. London: Mehtuen.

Judd, J.W. (ed.), 1890. *C.R. Darwin, On the Structure and Distribution of Coral Reefs also Geological Observations on the Volcanic Islands and Parts of South America Visited during the Voyage of H.M.S. 'Beagle'*. London: Ward, Lock & Co.

Jukes, J.B., 1862. On the mode of formation of some of the river-valleys in the South of Ireland. *Quarterly Journal, Geological Society, London*, **18**, 378–403.

Kennedy, B.A., 1976. Valley-side slopes and climate. In: E. Derbyshire (ed.), *Geomorphology and Climate*. London: Wiley, 171–201.

Kennedy, B.A., 1978. After Horton. *Earth Surface Processes*, **3**, 219–232.

Kennedy, B.A., 1979. A naughty world. *Transactions, Institute of British Geographers*, (NS) **4**, 550–558.

Kennedy, B.A., 1983. On outrageous hypotheses in geography. *Geography*, **68**, 326–330.

Kennedy, B.A., 1984. On Playfair's Law of accordant junctions. *Earth Surface Processes and Landforms*, **9**, 153–73.

Kennedy, B.A., 1992. Hutton to Horton: sequence, progress and equilibrium in geomorphology. *Geomorphology*, **5**, 231–50.

Kennedy, B.A., 1993. 'no prospect of an end . . .' *Geography*, **78**, 124–36.

Kennedy, B.A., 1997a. The trouble with valleys. In: D.R. Stoddart (ed.), *Process and Form in Geomorphology*. London: Routledge, 60–73.

Kennedy, B.A., 1997b. Classic papers revisited: Schumm and Lichty, 1965. *Progress in Physical Geography*, **21**, 419–23.

Kennedy, B.A., 2000. Trompe l'oeil. *Journal of Biogeography*, **27**, 37–38.

Kennedy, B.A., 2001. Charles Lyell and 'Modern changes of the Earth': the Milledgeville Gully. *Geomorphology*, **40**, 91–98.

Kennedy, B.A., 2004. Classic papers revisited. D.R. Stoddart, 1966. *Progress in Physical Geography*, **28**, 399–403.

Kerr, R., 2005. Titan, once a world apart, becomes eerily familiar. *Science*, **307**, 330–331.

Keylock, C., 2003. Mark Melton's geomorphology and geography's quantitative revolution. *Transactions, Institute of British Geographers* (NS), **28**, 142–57.

Keynes, R., 2002. *Fossils, Finches and Fuegians: Charles Darwin's Adventures and Discoveries on the Beagle, 1832–6*. London: Harper Collins.

King, L.C., 1951. *South African Scenery* (2nd edn). Edinburgh: Oliver & Boyd.

King, L.C., 1962. *Morphology of the Earth: a Study and Synthesis of World Scenery*. Edinburgh: Oliver and Boyd.

King, P.B. & Schumm, S.A., 1980. *The Physical Geography (Geomorphology) of William Morris Davis*. Norwich: Geo Books.

Kirkby, M.J., 1967. Measurement and theory of soil creep. *Journal of Geology*, **75**, 359–78.

Kirkby, M.J. & Chorley, R.J., 1967. Through flow, overland flow and erosion. *Bulletin, International Association of Scientific Hydrology*, **12**, 5–21.

Kirwan, R., 1799. *Geological Essays*. London: D. Bremner.

Knighton, A.D., 1998. *Fluvial Forms and Processes: a New Perspective* (2nd edn). London: Arnold.

Kuhn, T., 1962. *The Structure of Scientific Revolutions*. Chicago: Chicago University Press.

Lakatos, I., 1970. Falsification and the methodology of scientific research programmes. In: I. Lakatos & A. Musgrave (eds), *Criticism and the Growth of Knowledge*. Cambridge: Cambridge University Press, 91–196.

Lakatos, I., 1978. *The Methodology of Scientific Research Programmes*. Philosophical Papers (edited by J. Worrall & G. Currie), Vol. 1. Cambridge: Cambridge University Press.

Langbein, W.B. & Schumm, S.A., 1958. Yield of sediment in relation to mean annual precipitation. *American Geophysical Union, Transactions*, **39**, 1076–84.

Leigh, R.A. & Johnston, A.E. (eds), 1994. *Long-term Experiments in Agricultural and Ecological Sciences*. Wallingford: CAB International.

Leopold, L.B., 1994. *A View of the River*. Cambridge, MA: Harvard University Press.

Leopold, L.B. & Maddock, T., 1953. The hydraulic geometry of stream channels and some physiographic implications. *U.S. Geological Survey, Professional Paper* **252**.

Leopold, L.B. & Wolman, M.G., 1957. River channel patterns – braided, meandering and straight. *U.S. Geological Survey, Professional Paper*, **282B**, 39–85.

Leopold, L.B., Wolman, M.G. & Miller, J.P., 1964. *Fluvial Processes in Geomorphology*. San Francisco: W.H. Freeman & Co.

Livingstone, D.N., 1992. *The Geographical Tradition: Episodes in the History of a Contested Enterprise*. Oxford: Blackwell.

Lomborg, B., 2001. *The Skeptical Environmentalist: Measuring the Real State of the World*. Cambridge: Cambridge University Press.

Lovelock, J.E., 1979. *Gaia: a New Look at Life on Earth*. Oxford: Oxford University Press.

Lovelock, J.E., 1988. *The Ages of Gaia: a Biography of our Living Earth*. Oxford: Oxford University Press.

Lovelock, J.E., 1995. *The Ages of Gaia* (2nd edn). Oxford: Oxford University Press.

Lurie, E., 1960. *Louis Agassiz: a life in Science*. Chicago: University of Chicago Press.

Lyell, C., 1830–33. *Principles of Geology: Being an Attempt to Explain the Former Changes of the Earth's Surface, by Reference to Causes now in Operation*. London: John Murray (3 volumes).

Lyell, C., 1840. *Principles of Geology* (6th edn). London: John Murray (3 volumes).

Lyell, C., 1845. *Travels in North America*. London: John Murray.

Lyell, C., 1847. *Principles of Geology* (7th edn). London: John Murray.

Lyell, C., 1849. *A Second Visit to the United States of America*. London: John Murray (2 volumes).

Lyell, C., 1852. *A Manual of elementary Geology* (4th edn). London: John Murray.

Lyell, C., 1853. *Principles of Geology* (9th edn). London: John Murray.

Lyell, C., 1867. *Principles of Geology* (10th edn). Volume. London: John Murray.

Lyell, C., 1875. *Principles of Geology or the Modern Changes of the Earth and its Inhabitants Considered as Illustrative of Geology* (12th edn). London: John Murray (2 volumes).

MacCullough, J., 1814. On the granite tors of Cornwall. *Transactions, Geological Society of London*, **II**, 66–78.

MacCullough, J., 1816. A geological description of Glen Tilt. *Transactions, Geological Society of London*, **III**, 259–337.

Mackin, J.H., 1963. Rational and empirical methods of investigation in geology. In: C.C. Albritton (ed.), *The Fabric of Geology*. Stanford: Freeman, Cooper & Co., 135–63.

Malamud, B.D., Turcotte, D.L., Guzzetti, F. & Reichenbach, P., 2004. Landslide inventories and their statistical properties. *Earth Surface Processes and Landforms*, **29**, 687–711

Malin, M.C. & Edgett, K.S., 2003. Evidence for persistent flow and aqueous sedimentation on early Mars. *Science*, **302**, 1931–4.

Mandelbrot, B., 1967. How long is the coast of Britain? Statistical self-similarity and fractal dimensions. *Science*, **156**, 636–8.

Manning, R., 1890. On the flow of water in open channels and pipes. *Transactions, Institute of Civil Engineers, Ireland*, **20**, 161–207.

Masutti, C., 2002. L'héritage de Henri Baulig (1877–1962): éléments bibliographiques, bibliographie complète, inventaire des notes manuscrites. *Cybergeo*, **229**. 22 November 2002. http://www.cybergeo.presse.fr/ehgo/masutti/masut 02. htm.

Mayr, E., 1982. *The Growth of Biological Thought: Diversity, Evolution and Inheritance*. Cambridge, Mass: Belknap Press.

McIntyre, D.B. & McKirdy, A., 2001. *James Hutton: the Founder of Modern Geology* (revised edn). Edinburgh: National Museums of Scotland.

Melton, M.A., 1957. An analysis of the relationships among elements of climate, surface properties and geomorphology. *Office Naval Research Technical Report*, **11**.

Melton, M.A., 1958. Geometric properties of mature drainage systems and their representation in an E4 phase space. *Journal of Geology*, **66**, 25–54.

Melton, M.A., 1960. Intravalley variation in slope angles related to microclimate and erosional environment. *Geological Society of America, Bulletin*, **71**, 133–44.

Miller, A.J. & Gupta, A. (eds), 1999. *Varieties of Fluvial Form*. Chichester: John Wiley & Sons.

Nagel, E., 1961. *The Structure of Science: Problems in the Logic of Scientific Explanation*. London: Routledge & Kegan Paul.

Needham, J.N. (with Wang Ling), 1971. *Science and Civilization in China*, Vol. IV, Part 3, #28. Cambridge: Cambridge University Press.

Nogues, M.A.F., 1870. *Traité d'histoire Naturelle . . . 4 ème année: géologie appliquée*. Paris: Victor Masson & Fils.

Ollier, C., 1991. *Ancient Landforms*. London: Belhaven Press.

Oreskes, N., 1999. *The Rejection of Continental Drift: Theory and Method in American Earth Science*. Oxford: Oxford University Press.

Pallas, P.S., 1777–8. Observations sur la formation des montagnes et les changemens arrivé au Globe, particulièrement à l'égard de l'Empire de Russie. *Acta Academiae Scientiarum Imperialis Petropolitanae*, 21–64.

Pallas, P.S., 1802–3. (Translated.) *Travels through the Southern Provinces of the Russian Empire, in the Years 1793 and 1794*. London: Longman, Rees, T. Cadell Jun., Davies, Murray and Highley.

Penck, W., 1924. *Die Morphologische Analyse: Ein Kapitel der Physikalischen Geologie*. Stuttgart: J. Engelhorns Nachf. (Translated, 1953, by H. Czech & K.C. Boswell, *Morphological Analysis of Landforms*. London: Macmillan.)

Phillips, J.D., 1992. Qualitative chaos in geomorphic systems, with an example from wetland response to sea level rise. *Journal of Geology*, **100**, 365–74.

Piveteau, J. (ed.), 1954. *Oeuvres philosophiques de Buffon*. Paris: Presses Universitaires de France.

Playfair, J., 1802. *Illustrations of the Huttonian System of the Earth*. Edinburgh William Creech. (Reprinted in facsimile, 1956. New York: Dover Publications.)

Playfair, J., 1805. Biographical account of the late James Hutton, M.D., F.R.S. Edin. *Transactions, Royal Society of Edinburgh*, **V**, 41–99.

Popper, K.R., 1959. *The Logic of Scientific Discovery* (3rd edn). London: Hutchinson.

Powell, J.W., 1875. *Exploration of the Colorado River of the West (1869–72)*. Washington. (Reprinted, 1957, University of Chicago Press and Cambridge University Press.)

Powell, J.W., 1876. *Report on the Geology of the Eastern Portion of the Uinta Mountains*. Washington: Government Printer.

Powell, J.W., 1895. *Canyons of the Colorado*. (Reprinted, 1961. New York: Dover Publications.)

Prigogine, I., 1978. Time, structure and fluctuations. *Science*, **201**, pp. 777–85.

Pyne, S., 1980. *Grove Karl Gilbert: a Great Engine of Research*. Austin: University of Texas Press.

Ramsay, A.C., 1862. On the glacial origin of certain lakes in Switzerland, the Black Forest, Great Britain, Sweden, North America and elsewhere. *Quarterly Journal, Geological Society, London*, **18**, 185–204.

Ramsay, A.C., 1863. *The Physical Geology and Geography of Great Britain*. London: Stanford.

Rapp, A., 1960. Recent development of mountain slopes in Kärkevagge and surroundings, northern Scandinavia. *Geografiska Annaler*, **42**, 73–200.

Rapp, A., 1985. Extreme rainfall and rapid snowmelt as causes of mass movements in high latitude mountains, In: M.A. Church and O. Slaymaker (eds), *Field and Theory: Lectures in Geocryology*. Vancouver: University of British Columbia Press, 36–56.

Repchek, J., 2003. *The Man who found Time: James Hutton and the Discovery of the Earth's Antiquity*. Cambridge, Mass: Perseus Publishing.

Reusser, L.J., Bierman, P.R., Pavich, M.J., Zen, E-A., Larsen, J. & Finkel, R., 2004. Rapid Late Pleistocene incision of Atlantic passive-margin river gorges. *Science*, **305**, 499–502.

Rhoads, B.L. & C.E. Thorn (eds), 1996. *The Scientific Nature of Geomorphology*. Chichester: John Wiley.

Roberts, N., 1998. *The Holocene: an Environmental History* (2nd edn). Oxford: Blackwell.

Rodriguez-Iturbe, I. & Rinaldo, A., 1997. *Fractal River Basins: Change and Self-organization*. Cambridge: Cambridge University Press.

Roger, J., 1970. Buffon, Georges-Louis Leclerc, Comte de. In: *Dictionary of Scientific Biography* (C.C. Gillispie (ed. in chief)). New York: Scribner, Vol. II, 576–582.

Rouse, H. & Ince, S., 1957. *History of Hydraulics*. Dover: New York.

Rudwick, M., 1970. Introduction. In: facsimile edn, C. Lyell, *Principles of Geology*. Lehre: Verlag v.a. Cramer, Vol. 1, Ix–xxv.

Ruse, M., 1999. *Mystery of Mysteries: is Evolution a Social Construction?* Cambridge, Mass: Harvard University Press.

Saarinen, T.F., 1966. *Perception of the Drought Hazard in the Great Plains*. Department of Geography, University of Chicago, Research Paper 106.

Sack, D., 1992a. The trouble with antitheses: the case of G.K. Gilbert, geographer and educater. *Professional Geographer*, **43**, 28–37.

Sack, D., 1992b. New wine in old bottles: the Historiography of a paradigm change. *Geomorphology*, **5**, 251–63.

Saussure, H.B. de, 1779–1796. *Voyages dans les Alpes, précédés d'un Essai sur l'Histoire Naturelles des environs de Genève*. Neuchatel: L. Fauche-Borel (4 volumes).

Scheiddegger, A.E., 1961. *Theoretical Geomorphology*. Berlin: Springer-Verlag.

Scheiddegger, A.E., 1970. *Theoretical Geomorphology* (2nd edn). London: Allen & Unwin.

Scheiddegger, A.E., 1991. *Theoretical Geomorphology*. (3rd edn). Berlin: Springer-Verlag.

Schumm, S.A., 1956. The role of creep and rain-wash on the retreat of badland slopes. *American Journal of Science*, **254**, 693–706.

Schumm, S.A., 1977. *The Fluvial System*. New York: Wiley Interscience.

Schumm, S.A., 1991. *To Interpret the Earth: Ten Ways to be Wrong*. Cambridge: Cambridge University Press.

Schumm, S.A., 1997. Drainage density: problems of prediction and application, In: D.R. Stoddart (ed.), *Process and Form in Geomorphology*. London: Routledge, 15–45.

Schumm, S.A. & Chorley, R.J., 1964. The fall of Threatening Rock. *American Journal of Science*, **262**, 1041–54.

Schumm, S.A. & Lichty, R.W., 1965. Time, space, and causality in geomorphology. *American Journal of Science*, **263**, 110–19.

Scrope, G.J.P., 1825. *Considerations on Volcanoes, the Probable Causes of their Phenomena, Leading to the Establishment of a New Theory of the Earth*. London: W. Phillips.

Scrope, G.J.P., 1827. *Memoir on the Geology of Central France, including the Volcanic Formations of Auvergne, the Velay, and the Vivarais*. London: Longman, Rees, Orme, Brown & Green.

Scrope, G.J.P., 1834. (Read 5 February 1830.) On the formation of the valleys in which the Meuse, the Moselle, and some other rivers flow. *Proceedings, Geological Society of London*, **1**(4), 170–1.

Scrope, G.J.P., 1858. *The Geology and Extinct Volcanoes of Central France* (2nd edn of 1827 work). London: John Murray.

Secord, J.A. (ed.). *Charles Lyell: Principles of Geology* (1st edn). London: Penguin Books.

Selby, M.J., 1985. *Earth's Changing Surface*. Oxford: Clarendon Press.

Selby, M.J., 1993. *Hillslope Materials and Processes* (2nd edn). Oxford: Oxford University Press.

Shea, J.H., 1982. Twelve fallacies of Uniformitarianism. *Journal of Geological Education*, **10**, 455–60.

Shelton, J.S., 1966. *Geology Illustrated*. San Francisco: W.H. Freeman.

Shreve, R.L., 1966. Statistical law of stream numbers. *Journal of Geology*, **74**, 17–37.

Simpson, G.G., 1963. Historical Science. In: C.C. Albritton (ed.), *The Fabric of Geology*, Reading, Mass: Addison-Wesley, 24–48.

Sobel, D., 1995. *Longitude*. New York: Walker.

Sonnini, C.S. (ed.). An VII–1808 *Histoire Naturelle, générale et particulière, par LECLERC DE BUFFON . . . Ouvrage formant un cours complet d'histoire Naturelle*. Paris: Impriemérie F. Dufart (64 volumes).

Stoddart, D.R., 1966. Darwin's impact on Geography. *Annals, Association of American Geographers*, **56**, 683–98.

Stoddart, D.R., 1975. 'That Victorian Science': Huxley's *Physiography* and its impact on geography. *Transactions, Institute of British Geographers*, **66**, 17–40.

Stoddart, D.R., 1986. *On Geography and its History*. Oxford: Blackwells.

Stoddart, D.R. (ed.), 1997. *Process and Form in Geomorphology*. London: Routledge.

Stølum, H-H., 1996. River meandering as a self-organization process. *Science*, **271**, 1710–3.

Stott, R., 2003. *Darwin and the barnacle*. London: Faber & Faber.

Strahler, A.N., 1950. Equilibrium theory of erosional slopes approached by frequency distribution analysis. *American Journal of Science*, **248**, 673–96, 800–14.

Strahler, A.N., 1952a. Dynamic basis of geomorphology. *American Journal of Science*, **63**, 923–38.

Strahler, A.N., 1992. Quantitative/dynamic geomorphology at Columbia 1945–60: a retrospective. *Progress in Physical Geography*, **16**, 65–84.

Sundborg, A., 1956. The river Klaralven: a study in fluvial processes. *Geografiska Annaler*, **38**, 127–316.

Summerfield, M.A., 2005. The changing landscape of geomorphology. *Earth Surface Processes and Landforms*, **30**, 779–81.

Surell, A., 1841. *Etudes sur les torrents des Hautes-Alpes*. Paris: Carilian-Goeury and V. Dalmont.

Symons, D.T.A., 2004. Plate tectonics. In: A.S. Goudie (ed.), *Encyclopedia of Geomorphology*. London: Routledge, 792–6 (2 volumes).

Taylor, T.J., 1851. *An Inquiry into the Operation of Running Streams and Tidal Waters*. London: Longman, Brown, Green & Longmans.

Thorn, C.E. & M.R. Welford, 1994. The equilibrium concept in geomorphology. *Annals, Association of American Geographers*, **84**, 666–96 and discussion, 697–709.

Thornton, I., 1996. *Krakatau: the Destruction and Reassembly of an Island Ecosystem*. Cambridge, Mass: Harvard University Press.

Tinkler, K.J., 1985. *A short history of Geomorphology*. London: Croom Helm.

Tilley, P. (translator and editor), 1972. *A. Hettner: the Surface Features of the Land*. London: Macmillan.

Turcotte, D.L., 1997. *Fractals and Chaos in Geology and Geophysics* (2nd edn). Cambridge: Cambridge University Press.

Twain, M., 1883. *Life on the Mississippi*. Boston: Osgood.

Tyndall, J., 1860. *The Glaciers of the Alps*. London: John Murray.

Van der Lee, S., 2001. Deep below North America. *Science*, **294**, 1297–8.

Van Dover, C.L., 2000. *The Ecology of Deep Sea Hydrothermal Vents*. Princeton: Princeton University Press.

Viles, H.A. (ed.), 1988. *BioGeomorphology*. Oxford: Basil Blackwell.

Viles, H.A., 1995. Ecological perspectives on rock surface weathering: towards a conceptual model. *Geomorphology*, **13**, 21–35.

Walker, H.J. & Grabau, W.E. (eds), 1993. *The Evolution of Geomorphology: a Nation by Nation Summary of Development*. London: Wiley.

Wegener, A., 1924. *The Origin of Continents and Oceans*. English edition, 1967. London: Methuen.

Wilson, L.G., 1972. *Charles Lyell. The Years to 1841: the Revolution in Geology*. New Haven: Yale University Press.

Wilson, L.G., 1999. *Lyell in America: Transatlantic Geology, 1841–1853*. Baltimore: Johns Hopkins University Press.

Winchester, S., 2001. *The Map that Changed the World: the Tale of William Smith and the Birth of a Science*. London: Viking.

Winchester, S., 2003. *Krakatoa: the Day the World Exploded*. London: Harper Collins.

Woldenberg, M.J., 1968. Spatial order in fluvial systems: Horton's laws derived from mixed hexagonal hierarchies of drainage basin areas. *Geological Society of America, Bulletin*, **80**, 97–112.

Wolman, M.G. & Gerson, R., 1978. Relative scales of time and effectiveness of climate in watershed geomorphology. *Earth Surface Processes*, **3**, 189–208.

Wolman, M.G. & Miller, J.P., 1960. Magnitude and frequency of forces in geomorphic operation. *Journal of Geology*, **68**, 54–74.

Wooldridge, S.W., 1936. River profiles and denudation chronology in southern England. *Geological Magazine*, **73**, 1–16.

Worster, D., 2001. *A River Running West: the Life of John Wesley Powell*. Oxford: Oxford University Press.

Yochelson, E.L. (ed.), 1980. The scientific ideas of G.K. Gilbert: an assessment on the occasion of the centennial of the United States of Geological Survey (1879–1979). *U.S. Geological Survey Special Paper*, **183**.

Index